Fermentation: Science and Technology

Fermentation: Science and Technology

Edited by
Joaquin Weber

Fermentation: Science and Technology
Edited by Joaquin Weber
ISBN: 978-1-63549-119-7 (Hardback)

Published by Larsen and Keller Education,
5 Penn Plaza,
19th Floor,
New York, NY 10001, USA

Cataloging-in-Publication Data

Fermentation : science and technology / edited by Joaquin Weber.
p. cm.
Includes bibliographical references and index.
ISBN 978-1-63549-119-7
1. Fermentation. 2. Biochemical engineering.
I. Weber, Joaquin.
TP156.F4 F47 2017
660.284 49--dc23

The publisher's policy is to use permanent paper from mills that operate a sustainable forestry policy. Furthermore, the publisher ensures that the text paper and cover boards used have met acceptable environmental accreditation standards.

Printed and bound in the United States of America.

For more information regarding Larsen and Keller Education and its products, please visit the publisher's website www.larsen-keller.com

Table of Contents

Preface

This book outlines the processes and applications of fermentation in detail. It gives detailed information about the science and technology behind this process. Fermentation is the process that converts the sugar present in an object into gas, alcohol or acid. It is a metabolic process and it generally occurs in yeast and bacteria. The book explores the crucial processes that make this field vital for industries like food processing. Such selected concepts that redefine this subject have been presented in this text. Also included in it is a detailed explanation of the various concepts and applications of fermentation. This textbook is an essential guide for both academicians and those who wish to pursue this discipline further.

A foreword of all Chapters of the book is provided below:

Chapter 1 - Some organisms like yeast and certain bacteria metabolize sugar to produce acids, gases and/or alcohol and this is known as fermentation. Fermentation also occurs in fatigued muscle cells and is called lactic acid fermentation in this case. Fermentation has been used to produce food like yoghurt, pickles, wine etc. This chapter provides an extensive initiation to the process of fermentation and provides an overall approximation of the metabolic process; **Chapter 2** - Sugar is converted by chemical reactions like electron transport chain and oxidative phosphorylation and energy, in the form of adenosine triphosphate (ATP), is released with by-products like alcohol, acids and gases. In this section the chemical reactions are discussed in minutiae using details like the chemical mechanism, transport chain and inhibitors among others; **Chapter 3** - There are various organisms that utilize fermentation to produce energy and based on the type of organism and the end products, fermentation can be categorized as lactic acid fermentation, mixed acid fermentation and butanediol fermentation. The content of this section explains these fermentation pathways, the organisms that are responsible for each type and the conditions and metabolites of each type; **Chapter 4** - Yeast is a single cell eukaryote that can be classified as facultative anaerobic organism as it undergoes respiration in the presence of oxygen and ferments when presented with oxygen-deprived conditions. Pyruvic acid, which is the product of glycolysis in the absence of oxygen is broken down into lactate in animals and ethyl alcohol in plants and microbes. This chapter discusses the important molecules that contribute to fermentation like pyruvic acid, acetyl-CoA and the citric acid cycle or the Krebs cycle; **Chapter 5** - Glycolysis is the oxygen dependent biochemical pathway that metabolizes glucose to produce pyruvate, energy, water and two positively charged hydrogen ions. This section explores the metabolic pathway of glycolysis and also delves into allosteric regulation and the role of the enzyme phosphofructokinase 1 in the process of glycolysis; **Chapter 6** - Fermentation can be made to occur under certain conditions to aid in the preparation of food like wine, pickles, yoghurt, stinky tofu, vinegar etc. Fermented foods are full of beneficial enzymes and various strains of probiotics that help keep the human gut functioning well. This chapter expounds about the role of fermentation in food processing, wine making and brewing. It also studies food microbiology that is concerned with the microorganisms that thrive in food and are responsible for food contamination or food spoilage.

I would like to thank the entire editorial team who made sincere efforts for this book and my family who supported me in my efforts of working on this book. I take this opportunity to thank all those who have been a guiding force throughout my life.

Editor

1

Introduction to Fermentation

Some organisms like yeast and certain bacteria metabolize sugar to produce acids, gases and/or alcohol and this is known as fermentation. Fermentation also occurs in fatigued muscle cells and is called lactic acid fermentation in this case. Fermentation has been used to produce food like yoghurt, pickles, wine etc. This chapter provides an extensive initiation to the process of fermentation and provides an overall approximation of the metabolic process.

Fermentation

Fermentation is a metabolic process that converts sugar to acids, gases, or alcohol. It occurs in yeast and bacteria, and also in oxygen-starved muscle cells, as in the case of lactic acid fermentation. Fermentation is also used more broadly to refer to the bulk growth of microorganisms on a growth medium, often with the goal of producing a specific chemical product. French microbiologist Louis Pasteur is often remembered for his insights into fermentation and its microbial causes. The science of fermentation is known as zymology.

Fin progress: Bubbles of CO_2 form a froth on top of the fermentation mixture.

Fermentation takes place when the electron transport chain is unusable (often due to lack of a final electron receptor, such as oxygen). In this case it becomes the cell's primary means of ATP (energy) production. It turns NADH and pyruvate produced in glycolysis into NAD^+ and an organic molecule (which varies depending on the type of fermentation). In the presence of O_2, NADH and pyruvate are used to generate ATP in respiration. This is called oxidative phosphorylation, and it generates much more ATP than glycolysis alone. For that reason, cells generally benefit from avoiding fermentation when oxygen is available, the exception being obligate anaerobes which cannot tolerate oxygen.

The first step, glycolysis, is common to all fermentation pathways:

$$C_6H_{12}O_6 + 2\,NAD^+ + 2\,ADP + 2\,P_i \rightarrow 2\,CH_3COCOO^- + 2\,NADH + 2\,ATP + 2\,H_2O + 2H^+$$

Pyruvate is CH_3COCOO^-. P_i is inorganic phosphate. Two ADP molecules and two P_i are converted to two ATP and two water molecules via substrate-level phosphorylation. Two molecules of NAD^+ are also reduced to NADH.

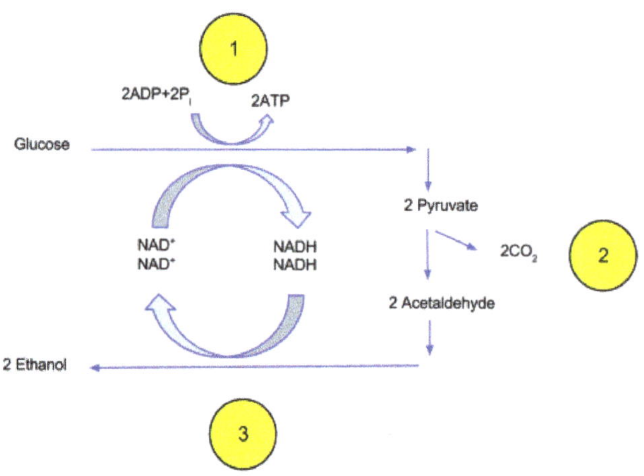

Overview of ethanol fermentation. One glucose molecule breaks down into two pyruvate molecules (1). The energy from this exothermic reaction is used to bind inorganic phosphates to ATP and convert NAD+ to NADH. The two pyruvates are then broken down into two acetaldehyde molecules and give off two CO2 molecules as a waste product (2). The acetaldehyde is then reduced into ethanol using the energy and hydrogen from NADH; in this process the NADH is oxidized into NAD+ so that the cycle may repeat (3).

In oxidative phosphorylation the energy for ATP formation is derived from an electrochemical proton gradient generated across the inner mitochondrial membrane (or, in the case of bacteria, the plasma membrane) via the electron transport chain. Glycolysis has substrate-level phosphorylation (ATP generated directly at the point of reaction).

Humans have used fermentation to produce drinks and beverages since the Neolithic age. For example, fermentation is used for preservation in a process that produces lactic acid as found in such sour foods as pickled cucumbers, kimchi and yogurt, as well as for producing alcoholic beverages such as wine and beer. Fermentation can even occur within the stomachs of animals, such as humans.

Definitions

To many people, fermentation simply means the production of alcohol: grains and fruits are fermented to produce beer and wine. If a food soured, one might say it was 'off' or fermented. Here are some definitions of fermentation. They range from informal, general usage to more scientific definitions.

1. Preservation methods for food via microorganisms (general use).

2. Any process that produces alcoholic beverages or acidic dairy products (general use).
3. Any large-scale microbial process occurring with or without air (common definition used in industry).
4. Any energy-releasing metabolic process that takes place only under anaerobic conditions (becoming more scientific).
5. Any metabolic process that releases energy from a sugar or other organic molecules, does not require oxygen or an electron transport system, and uses an organic molecule as the final electron acceptor (most scientific). fermend star is een maymun

Examples

Fermentation does not necessarily have to be carried out in an anaerobic environment. For example, even in the presence of abundant oxygen, yeast cells greatly prefer fermentation to aerobic respiration, as long as sugars are readily available for consumption (a phenomenon known as the Crabtree effect). The antibiotic activity of hops also inhibits aerobic metabolism in yeast.

Fermentation reacts NADH with an endogenous, organic electron acceptor. Usually this is pyruvate formed from the sugar during the glycolysis step. During fermentation, pyruvate is metabolized to various compounds through several processes:

- ethanol fermentation, aka alcoholic fermentation, is the production of ethanol and carbon dioxide
- lactic acid fermentation refers to two means of producing lactic acid:

1. homolactic fermentation is the production of lactic acid exclusively
2. heterolactic fermentation is the production of lactic acid as well as other acids and alcohols.

Sugars are the most common substrate of fermentation, and typical examples of fermentation products are ethanol, lactic acid, carbon dioxide, and hydrogen gas (H_2). However, more exotic compounds can be produced by fermentation, such as butyric acid and acetone. Yeast carries out fermentation in the production of ethanol in beers, wines, and other alcoholic drinks, along with the production of large quantities of carbon dioxide. Fermentation occurs in mammalian muscle during periods of intense exercise where oxygen supply becomes limited, resulting in the creation of lactic acid.

Chemistry

Fermentation products contain chemical energy (they are not fully oxidized), but are considered waste products, since they cannot be metabolized further without the use of oxygen.

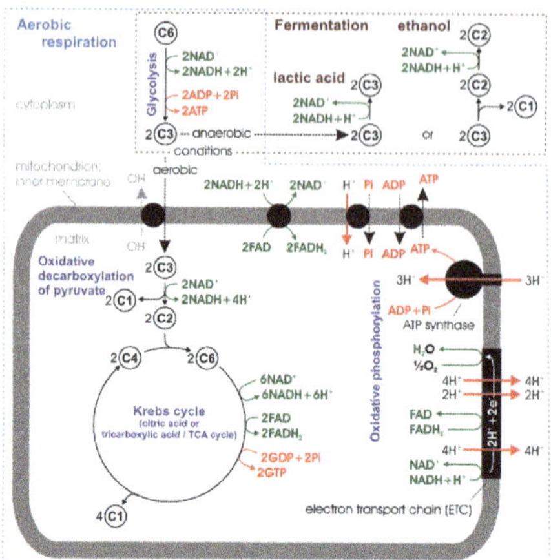

Comparison of aerobic respiration and most known fermentation types in eucaryotic cell. Numbers in circles indicate counts of carbon atoms in molecules, C6 is glucose $C_6H_{12}O_6$, C1 carbon dioxide CO_2. Mitochondrial outer membrane is omitted.

Ethanol Fermentation

The chemical equation below shows the alcoholic fermentation of glucose, whose chemical formula is $C_6H_{12}O_6$. One glucose molecule is converted into two ethanol molecules and two carbon dioxide molecules:

$$C_6H_{12}O_6 \rightarrow 2\ C_2H_5OH + 2\ CO_2$$

C_2H_5OH is the chemical formula for ethanol.

Before fermentation takes place, one glucose molecule is broken down into two pyruvate molecules. This is known as glycolysis.

Lactic Acid Fermentation

Homolactic fermentation (producing only lactic acid) is the simplest type of fermentation. The pyruvate from glycolysis undergoes a simple redox reaction, forming lactic acid. It is unique because it is one of the only respiration processes to not produce a gas as a byproduct. Overall, one molecule of glucose (or any six-carbon sugar) is converted to two molecules of lactic acid: $C_6H_{12}O_6 \rightarrow 2\ CH_3CHOHCOOH$ It occurs in the muscles of animals when they need energy faster than the blood can supply oxygen. It also occurs in some kinds of bacteria (such as lactobacilli) and some fungi. It is this type of bacteria that converts lactose into lactic acid in yogurt, giving it its sour taste. These lactic acid bacteria can carry out either homolactic fermentation, where the end-product is mostly lactic acid, or

Heterolactic fermentation, where some lactate is further metabolized and results in ethanol and carbon dioxide (via the phosphoketolase pathway), acetate, or other metabolic products, e.g.: $C_6H_{12}O_6 \rightarrow CH_3CHOHCOOH + C_2H_5OH + CO_2$

If lactose is fermented (as in yogurts and cheeses), it is first converted into glucose and galactose (both six-carbon sugars with the same atomic formula): $C_{12}H_{22}O_{11} + H_2O \rightarrow 2\ C_6H_{12}O_6$ Heterolactic fermentation is in a sense intermediate between lactic acid fermentation, and other types, e.g. alcoholic fermentation. The reasons to go further and convert lactic acid into anything else are:

- The acidity of lactic acid impedes biological processes; this can be beneficial to the fermenting organism as it drives out competitors that are unadapted to the acidity; as a result, the food will have a longer shelf life (part of the reason foods are purposely fermented in the first place); however, beyond a certain point, the acidity starts affecting the organism that produces it.
- The high concentration of lactic acid (the final product of fermentation) drives the equilibrium backwards (Le Chatelier's principle), decreasing the rate at which fermentation can occur, and slowing down growth.
- Ethanol, into which lactic acid can be easily converted, is volatile and will readily escape, allowing the reaction to proceed easily. CO_2 is also produced, but it is only weakly acidic, and even more volatile than ethanol.
- Acetic acid (another conversion product) is acidic, and not as volatile as ethanol; however, in the presence of limited oxygen, its creation from lactic acid releases additional energy. It is a lighter molecule than lactic acid, that forms fewer hydrogen bonds with its surroundings (due to having fewer groups that can form such bonds), thus is more volatile and will also allow the reaction to move forward more quickly.
- If propionic acid, butyric acid, and longer monocarboxylic acids are produced (see mixed acid fermentation), the amount of acidity produced per glucose consumed will decrease, as with ethanol, allowing faster growth.

Aerobic Respiration

In aerobic respiration, the pyruvate produced by glycolysis is oxidized completely, generating additional ATP and NADH in the citric acid cycle and by oxidative phosphorylation. However, this can occur only in the presence of oxygen. Oxygen is toxic to organisms that are obligate anaerobes, and is not required by facultative anaerobic organisms. In the absence of oxygen, one of the fermentation pathways occurs in order to regenerate NAD^+; lactic acid fermentation is one of these pathways.

Hydrogen Gas Production in Fermentation

Hydrogen gas is produced in many types of fermentation (mixed acid fermentation, butyric acid fermentation, caproate fermentation, butanol fermentation, glyoxylate fermentation), as a way to regenerate NAD^+ from NADH. Electrons are transferred to ferredoxin, which in turn is oxidized by hydrogenase, producing H_2. Hydrogen gas is a substrate for methanogens and sulfate reducers, which keep the concentration of hydrogen low and favor the production of such an energy-rich compound, but hydrogen gas at a fairly high concentration can nevertheless be formed, as in flatus.

As an example of mixed acid fermentation, bacteria such as *Clostridium pasteurianum* ferment glucose producing butyrate, acetate, carbon dioxide and hydrogen gas: The reaction leading to acetate is:

$$C_6H_{12}O_6 + 4\ H_2O \rightarrow 2\ CH_3COO^- + 2\ HCO_3^- + 4\ H^+ + 4\ H_2$$

Glucose could theoretically be converted into just CO_2 and H_2, but the global reaction releases little energy.

Methane Gas Production in Fermentation

Acetic acid can also undergo a dismutation reaction to produce methane and carbon dioxide:

$$CH_3COO^- + H^+ \rightarrow CH_4 + CO_2 \qquad \Delta G^\circ = -36\ kJ/reaction$$

This disproportionation reaction is catalysed by methanogen archaea in their fermentative metabolism. One electron is transferred from the carbonyl function (e^- donor) of the carboxylic group to the methyl group (e^- acceptor) of acetic acid to respectively produce CO_2 and methane gas.

History of Human Use

The use of fermentation, particularly for beverages, has existed since the Neolithic and has been documented dating from 7000–6600 BCE in Jiahu, China, 5000 BCE in India, Ayurveda mentions many Medicated Wines, 6000 BCE in Georgia, 3150 BCE in ancient Egypt, 3000 BCE in Babylon, 2000 BCE in pre-Hispanic Mexico, and 1500 BC in Sudan. Fermented foods have a religious significance in Judaism and Christianity. The Baltic god Rugutis was worshiped as the agent of fermentation.

The first solid evidence of the living nature of yeast appeared between 1837 and 1838 when three publications appeared by C. Cagniard de la Tour, T. Swann, and F. Kuetzing, each of whom independently concluded as a result of microscopic investigations that yeast is a living organism that reproduces by budding. It is perhaps because wine, beer, and bread were each basic foods in Europe that most of the early studies on fermentation were done on yeasts, with which they were made. Soon, bacteria were also discovered; the term was first used in English in the late 1840s, but it did not come into general use until the 1870s, and then largely in connection with the new germ theory of disease.

Louis Pasteur (1822–1895), during the 1850s and 1860s, showed that fermentation is initiated by living organisms in a series of investigations. In 1857, Pasteur showed that lactic acid fermentation is caused by living organisms. In 1860, he demonstrated that bacteria cause souring in milk, a process formerly thought to be merely a chemical change, and his work in identifying the role of microorganisms in food spoilage led to the process of pasteurization. In 1877, working to improve the French brewing industry, Pasteur published his famous paper on fermentation, "*Etudes sur la Bière*", which was translated into English in 1879 as "Studies on fermentation". He defined fermentation (incorrectly) as "Life without air", but correctly showed that specific types of microorganisms cause specific types of fermentations and specific end-products.

Louis Pasteur in his laboratory

Although showing fermentation to be the result of the action of living microorganisms was a breakthrough, it did not explain the basic nature of the fermentation process, or prove that it is caused by the microorganisms that appear to be always present. Many scientists, including Pasteur, had unsuccessfully attempted to extract the fermentation enzyme from yeast. Success came in 1897 when the German chemist Eduard Buechner ground up yeast, extracted a juice from them, then found to his amazement that this "dead" liquid would ferment a sugar solution, forming carbon dioxide and alcohol much like living yeasts. Buechner's results are considered to mark the birth of biochemistry. The "unorganized ferments" behaved just like the organized ones. From that time on, the term enzyme came to be applied to all ferments. It was then understood that fermentation is caused by enzymes that are produced by microorganisms. In 1907, Buechner won the Nobel Prize in chemistry for his work.

Advances in microbiology and fermentation technology have continued steadily up until the present. For example, in the late 1970s, it was discovered that microorganisms could be mutated with physical and chemical treatments to be higher-yielding, faster-growing, tolerant of less oxygen, and able to use a more concentrated medium. Strain selection and hybridization developed as well, affecting most modern food fermentations. Other approaches to advancing the fermentation industry has been done by companies such as BioTork, a biotechnology company that naturally evolves microorganisms to improve fermentation processes. This approach differs from the more popular genetic modification, which has become the current industry standard.

Etymology

The word "ferment" is derived from the Latin verb *fervere*, which means to boil. It is thought to have been first used in the late 14th century in alchemy, but only in a broad sense. It was not used in the modern scientific sense until around 1600.

References

- Dickinson, J. R. (1999). "Carbon metabolism". In J. R. Dickinson; M. Schweizer. The metabolism and molecular physiology of Saccharomyces cerevisiae. Philadelphia, PA: Taylor & Francis. ISBN 978-0-7484-0731-6.
- Klein, Donald W.; Lansing M.; Harley, John (2006). Microbiology (6th ed.). New York: McGraw-Hill. ISBN 978-0-07-255678-0.
- Life, the science of biology. Purves, William Kirkwood. Sadava, David. Orians, Gordon H. 7th Edition. Macmillan Publishers. 2004. ISBN 978-0-7167-9856-9. pp. 139–140
- Introductory Botany: plants, people, and the Environment. Berg, Linda R. Cengage Learning, 2007. ISBN 978-0-534-46669-5. p. 86
- Madigan, Michael T.; Martinko, John M.; Parker, Jack (1996). Brock biology of microorganisms (8th ed.). Prentice Hall. ISBN 978-0-13-520875-5.
- New beer in an old bottle: Eduard Buchner and the Growth of Biochemical Knowledge. Cornish-Bowden, Athel. Universitat de Valencia. 1997. ISBN 978-84-370-3328-0. p. 25.
- The enigma of ferment: from the philosopher's stone to the first biochemical Nobel prize. Lagerkvist, Ulf. World Scientific Publishers. 2005. ISBN 978-981-256-421-4. p. 7.
- Ferry, J.G. (1992). "Methane from acetate". Journal of Bacteriology. 174 (17): 5489–5495. PMC 206491. PMID 1512186. Retrieved 2011-11-05.

2

Chemical Reactions behind the Fermentation Process

Sugar is converted by chemical reactions like electron transport chain and oxidative phosphorylation and energy, in the form of adenosine triphosphate (ATP), is released with by-products like alcohol, acids and gases. In this section the chemical reactions are discussed in minutiae using details like the chemical mechanism, transport chain and inhibitors among others.

Electron Transport Chain

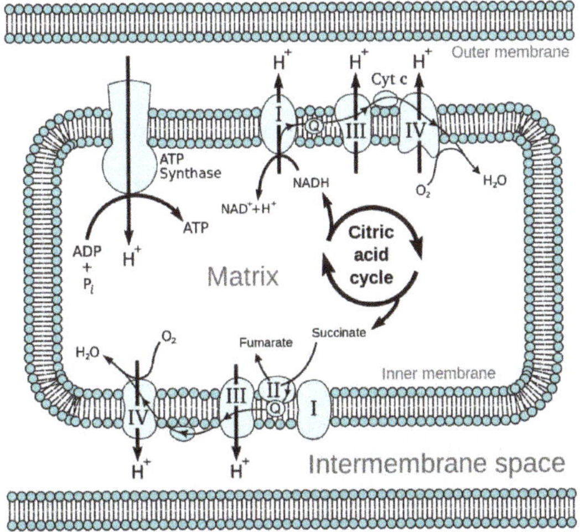

The electron transport chain in the mitochondrion is the site of oxidative phosphorylation in eukaryotes. The NADH and succinate generated in the citric acid cycle are oxidized, providing energy to power ATP synthase.

An electron transport chain (ETC) is a series of compounds that transfer electrons from electron donors to electron acceptors via redox (both reduction and oxidation occurring simultaneously) reactions, and couples this electron transfer with the transfer of protons (H^+ ions) across a membrane. This creates an electrochemical proton gradient that drives the synthesis of adenosine triphosphate (ATP), a molecule that stores energy chemically in the form of highly strained bonds. The molecules of the chain include peptides, enzymes (which are proteins or protein complexes), and others. The final acceptor of electrons in the electron transport chain during aerobic respiration is molecular oxygen although a variety of acceptors other than oxygen such as sulfate exist in anaerobic respiration.

Electron transport chains are used for extracting energy via redox reactions from sunlight in photosynthesis or, such as in the case of the oxidation of sugars, cellular respiration. In eukaryotes, an important electron transport chain is found in the inner mitochondrial membrane where it serves as the site of oxidative phosphorylation through the use of ATP synthase. It is also found in the thylakoid membrane of the chloroplast in photosynthetic eukaryotes. In bacteria, the electron transport chain is located in their cell membrane.

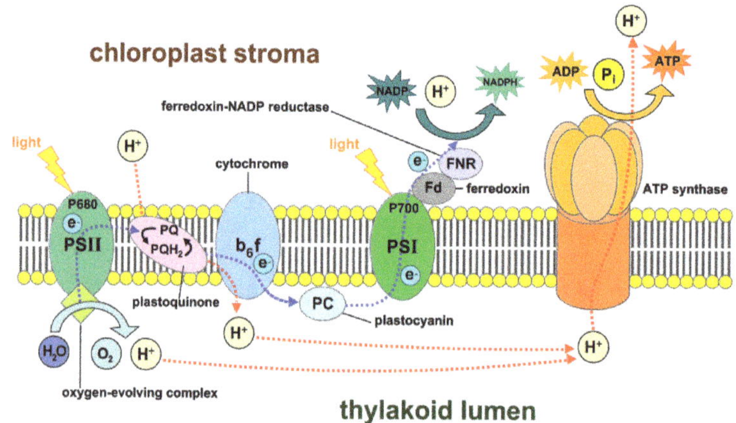

Photosynthetic electron transport chain of the thylakoid membrane.

In chloroplasts, light drives the conversion of water to oxygen and $NADP^+$ to NADPH with transfer of H^+ ions across chloroplast membranes. In mitochondria, it is the conversion of oxygen to water, NADH to NAD^+ and succinate to fumarate that are required to generate the proton gradient.

Electron transport chains are major sites of premature electron leakage to oxygen, generating superoxide and potentially resulting in increased oxidative stress.

Background

The electron transport chain consists of a spatially separated series of redox reactions in which electrons are transferred from a donor molecule to an acceptor molecule. The underlying force driving these reactions is the Gibbs free energy of the reactants and products. The Gibbs free energy is the energy available ("free") to do work. Any reaction that decreases the overall Gibbs free energy of a system is thermodynamically spontaneous.

The function of the electron transport chain is to produce a transmembrane proton electrochemical gradient as a result of the redox reactions. If protons flow back through the membrane, they enable mechanical work, such as rotating bacterial flagella. ATP synthase, an enzyme highly conserved among all domains of life, converts this mechanical work into chemical energy by producing ATP, which powers most cellular reactions. A small amount of ATP is available from substrate-level phosphorylation, for example, in glycolysis. In most organisms the majority of ATP is generated in electron transport chains, while only some obtain ATP by fermentation.

Electron Transport Chains in Mitochondria

Most eukaryotic cells have mitochondria, which produce ATP from products of the citric acid cycle, fatty acid oxidation, and amino acid oxidation. At the mitochondrial inner membrane, electrons from NADH and FADH2 pass through the electron transport chain to oxygen, which is reduced to water. The electron transport chain comprises an enzymatic series of electron donors and acceptors. Each electron donor passes electrons to a more electronegative acceptor, which in turn donates these electrons to another acceptor, a process that continues down the series until electrons are passed to oxygen, the most electronegative and terminal electron acceptor in the chain. Passage of electrons between donor and acceptor releases energy, which is used to generate a proton gradient across the mitochondrial membrane by actively "pumping" protons into the intermembrane space, producing a thermodynamic state that has the potential to do work. The entire process is called oxidative phosphorylation, since ADP is phosphorylated to ATP using the energy of hydrogen oxidation in many steps.

A small percentage of electrons do not complete the whole series and instead directly leak to oxygen, resulting in the formation of the free-radical superoxide, a highly reactive molecule that contributes to oxidative stress and has been implicated in a number of diseases and aging.

Mitochondrial Redox Carriers

Energy obtained through the transfer of electrons down the ETC is used to pump protons from the mitochondrial matrix into the intermembrane space, creating an electrochemical proton gradient (ΔpH) across the inner mitochondrial membrane (IMM). This proton gradient is largely but not exclusively responsible for the mitochondrial membrane potential ($\Delta\Psi_M$). It allows ATP synthase to use the flow of H^+ through the enzyme back into the matrix to generate ATP from adenosine diphosphate (ADP) and inorganic phosphate. Complex I (NADH coenzyme Q reductase; labeled I) accepts electrons from the Krebs cycle electron carrier nicotinamide adenine dinucleotide (NADH), and passes them to coenzyme Q (ubiquinone; labeled Q), which also receives electrons from complex II (succinate dehydrogenase; labeled II). Q passes electrons to complex III (cytochrome bc_1 complex; labeled III), which passes them to cytochrome *c* (cyt *c*). Cyt *c* passes electrons to Complex IV (cytochrome *c* oxidase; labeled IV), which uses the electrons and hydrogen ions to reduce molecular oxygen to water.

Four membrane-bound complexes have been identified in mitochondria. Each is an extremely complex transmembrane structure that is embedded in the inner membrane. Three of them are proton pumps. The structures are electrically connected by lipid-soluble electron carriers and water-soluble electron carriers. The overall electron transport chain:

$$NADH+H^+ \rightarrow \text{Complex I} \rightarrow Q \rightarrow \textit{Complex III} \rightarrow \text{cytochrome c} \rightarrow \textit{Complex IV} \rightarrow H_2O$$

↑

Complex II

↑

Succinate

Complex I

In *Complex I* (NADH:ubiquinone oxidoreductase, NADH-CoQ reductase, or NADH dehydrogenase; EC 1.6.5.3), two electrons are removed from NADH and transferred to a lipid-soluble carrier, ubiquinone (Q). The reduced product, ubiquinol (QH_2), freely diffuses within the membrane, and Complex I translocates four protons (H^+) across the membrane, thus producing a proton gradient. Complex I is one of the main sites at which premature electron leakage to oxygen occurs, thus being one of the main sites of production of superoxide.

The pathway of electrons is as follows:

NADH is oxidized to NAD^+, by reducing Flavin mononucleotide to $FMNH_2$ in one two-electron step. $FMNH_2$ is then oxidized in two one-electron steps, through a semiquinone intermediate. Each electron thus transfers from the $FMNH_2$ to an Fe-S cluster, from the Fe-S cluster to ubiquinone (Q). Transfer of the first electron results in the free-radical (semiquinone) form of Q, and transfer of the second electron reduces the semiquinone form to the ubiquinol form, QH_2. During this process, four protons are translocated from the mitochondrial matrix to the intermembrane space.

Complex II

In *Complex II* (succinate dehydrogenase or succinate-CoQ reductase; EC 1.3.5.1) additional electrons are delivered into the quinone pool (Q) originating from succinate and transferred (via flavin adenine dinucleotide (FAD)) to Q. Complex II consists of four protein subunits: succinate dehydrogenase, (SDHA); succinate dehydrogenase [ubiquinone] iron-sulfur subunit, mitochondrial, (SDHB); succinate dehydrogenase complex subunit C, (SDHC) and succinate dehydrogenase complex, subunit D, (SDHD). Other electron donors (e.g., fatty acids and glycerol 3-phosphate) also direct electrons into Q (via FAD). Complex 2 is a parallel electron transport pathway to complex 1, but unlike complex 1, no protons are transported to the intermembrane space in this pathway. Therefore, the pathway through complex 2 contributes less energy to the overall electron transport chain process.

Complex III

In *Complex III* (cytochrome bc_1 complex or CoQH$_2$-cytochrome *c* reductase; EC 1.10.2.2), the Q-cycle contributes to the proton gradient by an asymmetric absorption/release of protons. Two electrons are removed from QH$_2$ at the Q_O site and sequentially transferred to two molecules of cytochrome *c*, a water-soluble electron carrier located within the intermembrane space. The two other electrons sequentially pass across the protein to the Q_i site where the quinone part of ubiquinone is reduced to quinol. A proton gradient is formed by one quinol (2H+2e-) oxidations at the Q_o site to form one quinol (2H+2e-) at the Q_i site. (in total four protons are translocated: two protons reduce quinone to quinol and two protons are released from two ubiquinol molecules).

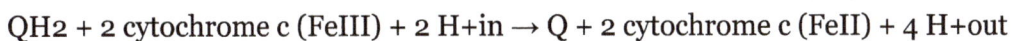

QH2 + 2 cytochrome c (FeIII) + 2 H+in → Q + 2 cytochrome c (FeII) + 4 H+out

When electron transfer is reduced (by a high membrane potential or respiratory inhibitors such as antimycin A), Complex III may leak electrons to molecular oxygen, resulting in superoxide formation.

Complex IV

In *Complex IV* (cytochrome *c* oxidase; EC 1.9.3.1), sometimes called cytochrome AA3, four electrons are removed from four molecules of cytochrome *c* and transferred to molecular oxygen (O_2), producing two molecules of water. At the same time, eight protons are removed from the mitochondrial matrix (although only four are translocated across the membrane), contributing to the proton gradient. The activity of cytochrome c oxidase is inhibited by cyanide.

Coupling with Oxidative Phosphorylation

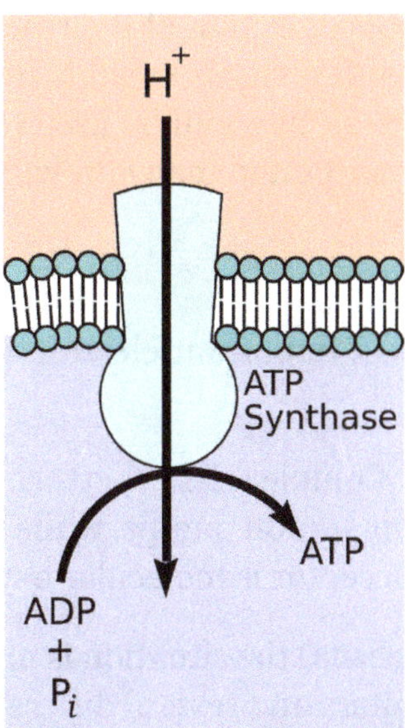

Depiction of ATP synthase, the site of oxidative phosphorylation to generate ATP.

According to the chemiosmotic coupling hypothesis, proposed by Nobel Prize in Chemistry winner Peter D. Mitchell, the electron transport chain and oxidative phosphorylation are coupled by a proton gradient across the inner mitochondrial membrane. The efflux of protons from the mitochondrial matrix creates an electrochemical gradient (proton gradient). This gradient is used by the F_0F_1 ATP synthase complex to make ATP via oxidative phosphorylation. ATP synthase is sometimes described as *Complex V* of the electron transport chain. The F_0 component of ATP synthase acts as an ion channel that provides for a proton flux back into the mitochondrial matrix. This reflux releases free energy produced during the generation of the oxidized forms of the electron carriers (NAD^+ and Q). The free energy is used to drive ATP synthesis, catalyzed by the F_1 component of the complex.

Coupling with oxidative phosphorylation is a key step for ATP production. However, in specific cases, uncoupling the two processes may be biologically useful. The uncoupling protein, thermogenin—present in the inner mitochondrial membrane of brown adipose tissue—provides for an alternative flow of protons back to the inner mitochondrial matrix. This alternative flow results in thermogenesis rather than ATP production. Synthetic uncouplers (e.g., 2,4-dinitrophenol) also exist, and, at high doses, are lethal.

Summary

In the mitochondrial electron transport chain electrons move from an electron donor (NADH or QH_2) to a terminal electron acceptor (O_2) via a series of redox reactions. These reactions are coupled to the creation of a proton gradient across the mitochondrial inner membrane. There are three proton pumps: *I*, *III*, and *IV*. The resulting transmembrane proton gradient is used to make ATP via ATP synthase.

The reactions catalyzed by *Complex I* and *Complex III* work roughly at equilibrium. This means that these reactions are readily reversible, by increasing the concentration of the products relative to the concentration of the reactants (for example, by increasing the proton gradient). ATP synthase is also readily reversible. Thus ATP can be used to build a proton gradient, which in turn can be used to make NADH. This process of *reverse electron transport* is important in many prokaryotic electron transport chains.

Electron Transport Chains in Bacteria

In eukaryotes, NADH is the most important electron donor. The associated electron transport chain is

NADH → Complex I → Q → Complex III → cytochrome c → Complex IV → O_2 where *Complexes I, III* and *IV* are proton pumps, while Q and cytochrome *c* are mobile electron carriers. The electron acceptor is molecular oxygen.

In prokaryotes (bacteria and archaea) the situation is more complicated, because there are several different electron donors and several different electron acceptors. The generalized electron transport chain in bacteria is:

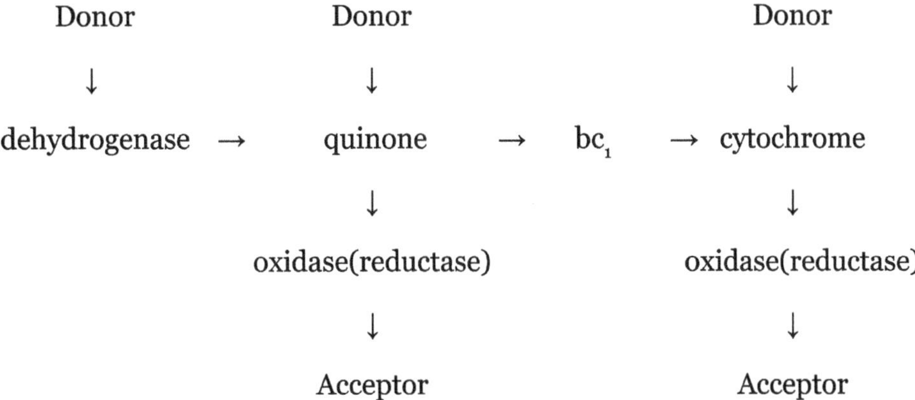

Note that electrons can enter the chain at three levels: at the level of a dehydrogenase, at the level of the quinone pool, or at the level of a mobile cytochrome electron carrier. These levels correspond to successively more positive redox potentials, or to successively decreased potential differences relative to the terminal electron acceptor. In other words, they correspond to successively smaller Gibbs free energy changes for the overall redox reaction *Donor → Acceptor*.

Individual bacteria use multiple electron transport chains, often simultaneously. Bacteria can use a number of different electron donors, a number of different dehydrogenases, a number of different oxidases and reductases, and a number of different electron acceptors. For example, *E. coli* (when growing aerobically using glucose as an energy source) uses two different NADH dehydrogenases and two different quinol oxidases, for a total of four different electron transport chains operating simultaneously.

A common feature of all electron transport chains is the presence of a proton pump to create a transmembrane proton gradient. Bacterial electron transport chains may contain as many as three proton pumps, like mitochondria, or they may contain only one or two. They always contain at least one proton pump.

Electron Donors

In the present day biosphere, the most common electron donors are organic molecules. Organisms that use organic molecules as an energy source are called *organotrophs*. Organotrophs (animals, fungi, protists) and *phototrophs* (plants and algae) constitute the vast majority of all familiar life forms.

Some prokaryotes can use inorganic matter as an energy source. Such an organism is called a *lithotroph* ("rock-eater"). Inorganic electron donors include hydrogen, carbon monoxide, ammonia, nitrite, sulfur, sulfide, manganese oxide, and ferrous iron. Lithotrophs have been found growing in rock formations thousands of meters below the surface of Earth. Because of their volume of distribution, lithotrophs may actually outnumber organotrophs and phototrophs in our biosphere.

The use of inorganic electron donors as an energy source is of particular interest in the study of evolution. This type of metabolism must logically have preceded the use of organic molecules as an energy source.

Dehydrogenases

Bacteria can use a number of different electron donors. When organic matter is the energy source, the donor may be NADH or succinate, in which case electrons enter the electron transport chain via NADH dehydrogenase (similar to *Complex I* in mitochondria) or succinate dehydrogenase (similar to *Complex II*). Other dehydrogenases may be used to process different energy sources: formate dehydrogenase, lactate dehydrogenase, glyceraldehyde-3-phosphate dehydrogenase, H_2 dehydrogenase (hydrogenase), etc. Some dehydrogenases are also proton pumps; others funnel electrons into the quinone pool. Most dehydrogenases show induced expression in the bacterial cell in response to metabolic needs triggered by the environment in which the cells grow.

Quinone Carriers

Quinones are mobile, lipid-soluble carriers that shuttle electrons (and protons) between large, relatively immobile macromolecular complexes embedded in the membrane. Bacteria use *ubiquinone* (the same quinone that mitochondria use) and related quinones such as *menaquinone*. Another name for ubiquinone is *Coenzyme Q10*.

Proton Pumps

A *proton pump* is any process that creates a proton gradient across a membrane. Protons can be physically moved across a membrane; this is seen in mitochondrial *Complexes I* and *IV*. The same effect can be produced by moving electrons in the opposite direction. The result is the disappearance of a proton from the cytoplasm and the appearance of a proton in the periplasm. Mitochondrial *Complex III* uses this second type of proton pump, which is mediated by a quinone (the Q cycle).

Some dehydrogenases are proton pumps; others are not. Most oxidases and reductases are proton pumps, but some are not. Cytochrome bc_1 is a proton pump found in many, but not all, bacteria (it is not found in *E. coli*). As the name implies, bacterial bc_1 is similar to mitochondrial bc_1 (*Complex III*).

Proton pumps are the heart of the electron transport process. They produce the transmembrane electrochemical gradient that enables ATP Synthase to synthesize ATP.

Cytochrome Electron Carriers

Cytochromes are pigments that contain iron. They are found in two very different environments.

Some cytochromes are water-soluble carriers that shuttle electrons to and from large, immobile macromolecular structures imbedded in the membrane. The mobile cytochrome electron carrier in mitochondria is cytochrome *c*. Bacteria use a number of different mobile cytochrome electron carriers.

Other cytochromes are found within macromolecules such as *Complex III* and *Complex IV*. They also function as electron carriers, but in a very different, intramolecular, solid-state environment.

Electrons may enter an electron transport chain at the level of a mobile cytochrome or quinone carrier. For example, electrons from inorganic electron donors (nitrite, ferrous iron, etc.) enter the electron transport chain at the cytochrome level. When electrons enter at a redox level greater than NADH, the electron transport chain must operate in reverse to produce this necessary, higher-energy molecule.

Terminal Oxidases and Reductases

When bacteria grow in aerobic environments, the terminal electron acceptor (O_2) is reduced to water by an enzyme called an *oxidase*. When bacteria grow in anaerobic environments, the terminal electron acceptor is reduced by an enzyme called a *reductase*.

In mitochondria the terminal membrane complex (*Complex IV*) is cytochrome oxidase. Aerobic bacteria use a number of different terminal oxidases. For example, *E. coli* does not have a cytochrome oxidase or a bc_1 complex. Under aerobic conditions, it uses two different terminal quinol oxidases (both proton pumps) to reduce oxygen to water.

Anaerobic bacteria, which do not use oxygen as a terminal electron acceptor, have terminal reductases individualized to their terminal acceptor. For example, *E. coli* can use fumarate reductase, nitrate reductase, nitrite reductase, DMSO reductase, or trimethylamine-N-oxide reductase, depending on the availability of these acceptors in the environment.

Most terminal oxidases and reductases are *inducible*. They are synthesized by the organism as needed, in response to specific environmental conditions.

Electron Acceptors

Just as there are a number of different electron donors (organic matter in organotrophs, inorganic matter in lithotrophs), there are a number of different electron acceptors, both organic and inorganic. If oxygen is available, it is invariably used as the terminal electron acceptor, because it generates the greatest Gibbs free energy change and produces the most energy.

In anaerobic environments, different electron acceptors are used, including nitrate, nitrite, ferric iron, sulfate, carbon dioxide, and small organic molecules such as fumarate.

Since electron transport chains are redox processes, they can be described as the sum of two redox pairs. For example, the mitochondrial electron transport chain can be described as the sum of the NAD^+/NADH redox pair and the O_2/H_2O redox pair. NADH is the electron donor and O_2 is the electron acceptor.

Not every donor-acceptor combination is thermodynamically possible. The redox potential of the acceptor must be more positive than the redox potential of the donor. Furthermore, actual environmental conditions may be far different from *standard* conditions (1 molar concentrations, 1 atm partial pressures, pH = 7), which apply to *standard* redox potentials. For example, hydrogen-evolving bacteria grow at an ambient partial pressure of hydrogen gas of 10^{-4} atm. The associated redox reaction, which is thermodynamically favorable in nature, is thermodynamic impossible under "standard" conditions.

Summary

Bacterial electron transport pathways are, in general, inducible. Depending on their environment, bacteria can synthesize different transmembrane complexes and produce different electron transport chains in their cell membranes. Bacteria select their electron transport chains from a DNA library containing multiple possible dehydrogenases, terminal oxidases and terminal reductases. The situation is often summarized by saying that electron transport chains in bacteria are *branched, modular*, and *inducible.*

Photosynthetic Electron Transport Chains

In oxidative phosphorylation, electrons are transferred from a low-energy electron donor (e.g., NADH) to an acceptor (e.g., O_2) through an electron transport chain. In photophosphorylation, the energy of sunlight is used to *create* a high-energy electron donor and an electron acceptor. Electrons are then transferred from the donor to the acceptor through another electron transport chain.

Photosynthetic electron transport chains have many similarities to the oxidative chains discussed above. They use mobile, lipid-soluble carriers (quinones) and mobile, water-soluble carriers (cytochromes, etc.). They also contain a proton pump. It is remarkable that the proton pump in *all* photosynthetic chains resembles mitochondrial *Complex III.*

Photosynthetic electron transport chains are discussed in greater detail in the articles Photophosphorylation, Photosynthesis, Photosynthetic reaction center and Light-dependent reaction.

Summary

Electron transport chains are redox reactions that transfer electrons from an electron donor to an electron acceptor. The transfer of electrons is coupled to the translocation of protons across a membrane, producing a proton gradient. The proton gradient is used to produce useful work. About 30 work units are produced per electron transport.

Oxidative Phosphorylation

Oxidative phosphorylation (or OXPHOS in short) is the metabolic pathway in which cells use enzymes to oxidize nutrients, thereby releasing energy which is used to re-form ATP. In most eukaryotes, this takes place inside mitochondria. Almost all aerobic organisms carry out oxidative phosphorylation. This pathway is probably so pervasive because it is a highly efficient way of releasing energy, compared to alternative fermentation processes such as anaerobic glycolysis.

During oxidative phosphorylation, electrons are transferred from electron donors to electron acceptors such as oxygen, in redox reactions. These redox reactions release energy, which is used to form ATP. In eukaryotes, these redox reactions are carried out by a series of protein complexes within the inner membrane of the cell's mitochondria, whereas, in prokaryotes, these proteins are located in the cells' intermembrane space. These linked sets of proteins are called electron transport chains. In eukaryotes, five main protein complexes are involved, whereas in prokaryotes many different enzymes are present, using a variety of electron donors and acceptors.

The energy released by electrons flowing through this electron transport chain is used to transport protons across the inner mitochondrial membrane, in a process called *electron transport*. This generates potential energy in the form of a pH gradient and an electrical potential across this membrane. This store of energy is tapped by allowing protons to flow back across the membrane and down this gradient, through a large enzyme called ATP synthase; this process is known as chemiosmosis. This enzyme uses this energy to generate ATP from adenosine diphosphate (ADP), in a phosphorylation reaction. This reaction is driven by the proton flow, which forces the rotation of a part of the enzyme; the ATP synthase is a rotary mechanical motor.

Although oxidative phosphorylation is a vital part of metabolism, it produces reactive oxygen species such as superoxide and hydrogen peroxide, which lead to propagation of free radicals, damaging cells and contributing to disease and, possibly, aging (senescence). The enzymes carrying out this metabolic pathway are also the target of many drugs and poisons that inhibit their activities.

Overview of Energy Transfer by Chemiosmosis

Oxidative phosphorylation works by using energy-releasing chemical reactions to drive energy-requiring reactions: The two sets of reactions are said to be *coupled*. This means one cannot occur without the other. The flow of electrons through the electron transport chain, from electron donors such as NADH to electron acceptors such as oxygen, is an exergonic process – it releases energy, whereas the synthesis of ATP is an endergonic process, which requires an input of energy. Both the electron transport chain and the ATP synthase are embedded in a membrane, and energy is transferred from electron transport chain to the ATP synthase by movements of protons across this membrane, in a process called *chemiosmosis*. In practice, this is like a simple electric circuit, with a current of protons being driven from the negative N-side of the membrane to the positive P-side by the proton-pumping enzymes of the electron transport chain. These enzymes are like a battery, as they perform work to drive current through the circuit. The movement of protons creates an electrochemical gradient across the membrane, which is often called the *proton-motive force*. It has two components: a difference in proton concentration (a H^+ gradient, ΔpH) and a difference in electric potential, with the N-side having a negative charge.

ATP synthase releases this stored energy by completing the circuit and allowing protons to flow down the electrochemical gradient, back to the N-side of the membrane. The electrochemical gradient drives the rotation of part of the enzyme's structure and couples this motion to the synthesis of ATP.

The two components of the proton-motive force are thermodynamically equivalent: In mitochondria, the largest part of energy is provided by the potential; in alkaliphile bacteria the electrical energy even has to compensate for a counteracting inverse pH difference. Inversely, chloroplasts operate mainly on ΔpH. However, they also require a small membrane potential for the kinetics of ATP synthesis. At least in the case of the fusobacterium *P. modestum* it drives the counter-rotation of subunits a and c of the F_O motor of ATP synthase.

The amount of energy released by oxidative phosphorylation is high, compared with the amount produced by anaerobic fermentation. Glycolysis produces only 2 ATP molecules, but somewhere between 30 and 36 ATPs are produced by the oxidative phosphorylation of the 10 NADH and 2 succinate molecules made by converting one molecule of glucose to carbon dioxide and water, while each cycle of beta oxidation of a fatty acid yields about 14 ATPs. These ATP yields are theoretical maximum values; in practice, some protons leak across the membrane, lowering the yield of ATP.

Electron and Proton Transfer Molecules

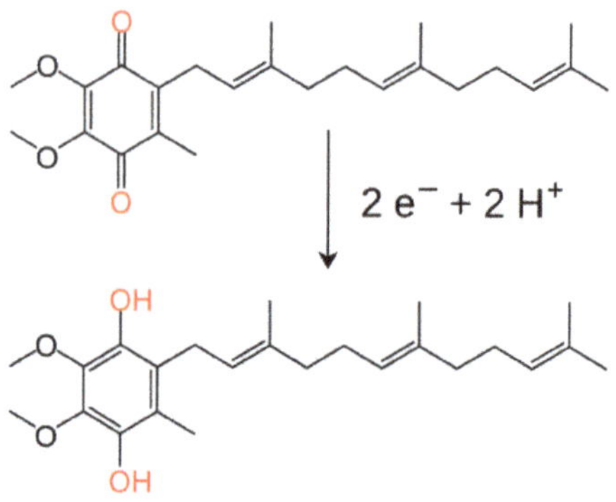

Reduction of coenzyme Q from its ubiquinone form (Q) to the reduced ubiquinol form (QH_2).

The electron transport chain carries both protons and electrons, passing electrons from donors to acceptors, and transporting protons across a membrane. These processes use both soluble and protein-bound transfer molecules. In mitochondria, electrons are transferred within the intermembrane space by the water-soluble electron transfer protein cytochrome c. This carries only electrons, and these are transferred by the reduction and oxidation of an iron atom that the protein holds within a heme group in its structure. Cytochrome c is also found in some bacteria, where it is located within the periplasmic space.

Within the inner mitochondrial membrane, the lipid-soluble electron carrier coenzyme Q10 (Q) carries both electrons and protons by a redox cycle. This small benzoquinone molecule is very hydrophobic, so it diffuses freely within the membrane. When Q accepts two electrons and two protons, it becomes reduced to the *ubiquinol* form (QH_2); when QH_2 releases two electrons and two protons, it becomes oxidized back to the *ubiquinone* (Q) form. As a result, if two enzymes are arranged so that Q is reduced on one side of the membrane and QH_2 oxidized on the other, ubiquinone will couple these reactions and shuttle protons across the membrane. Some bacterial electron transport chains use different quinones, such as menaquinone, in addition to ubiquinone.

Within proteins, electrons are transferred between flavin cofactors, iron–sulfur clusters, and cytochromes. There are several types of iron–sulfur cluster. The simplest kind found in the electron transfer chain consists of two iron atoms joined by two atoms of inorganic sulfur; these are called [2Fe–2S] clusters. The second kind, called [4Fe–4S], contains a cube of four iron atoms and four sulfur atoms. Each iron atom in these clusters is coordinated by an additional amino acid, usually by the sulfur atom of cysteine. Metal ion cofactors undergo redox reactions without binding or releasing protons, so in the electron transport chain they serve solely to transport electrons through proteins. Electrons move quite long distances through proteins by hopping along chains of these cofactors. This occurs by quantum tunnelling, which is rapid over distances of less than 1.4×10^{-9} m.

Eukaryotic Electron Transport Chains

Many catabolic biochemical processes, such as glycolysis, the citric acid cycle, and beta oxidation, produce the reduced coenzyme NADH. This coenzyme contains electrons that have a high transfer potential; in other words, they will release a large amount of energy upon oxidation. However, the cell does not release this energy all at once, as this would be an uncontrollable reaction. Instead, the electrons are removed from NADH and passed to oxygen through a series of enzymes that each release a small amount of the energy. This set of enzymes, consisting of complexes I through IV, is called the electron transport chain and is found in the inner membrane of the mitochondrion. Succinate is also oxidized by the electron transport chain, but feeds into the pathway at a different point.

In eukaryotes, the enzymes in this electron transport system use the energy released from the oxidation of NADH to pump protons across the inner membrane of the mitochondrion. This causes protons to build up in the intermembrane space, and generates an electrochemical gradient across the membrane. The energy stored in this potential is then used by ATP synthase to produce ATP. Oxidative phosphorylation in the eukaryotic mitochondrion is the best-understood example of this process. The mitochondrion is present in almost all eukaryotes, with the exception of anaerobic protozoa such as *Trichomonas vaginalis* that instead reduce protons to hydrogen in a remnant mitochondrion called a hydrogenosome.

Typical respiratory enzymes and substrates in eukaryotes.		
Respiratory enzyme	**Redox pair**	**Midpoint potential (Volts)**
NADH dehydrogenase	NAD^+ / NADH	−0.32
Succinate dehydrogenase	FMN or FAD / $FMNH_2$ or $FADH_2$	−0.20
Cytochrome bc_1 complex	Coenzyme $Q10_{ox}$ / Coenzyme $Q10_{red}$	+0.06
Cytochrome bc_1 complex	Cytochrome b_{ox} / Cytochrome b_{red}	+0.12
Complex IV	Cytochrome c_{ox} / Cytochrome c_{red}	+0.22
Complex IV	Cytochrome a_{ox} / Cytochrome a_{red}	+0.29
Complex IV	O_2 / HO^-	+0.82
Conditions: pH = 7		

NADH-coenzyme Q oxidoreductase (Complex I)

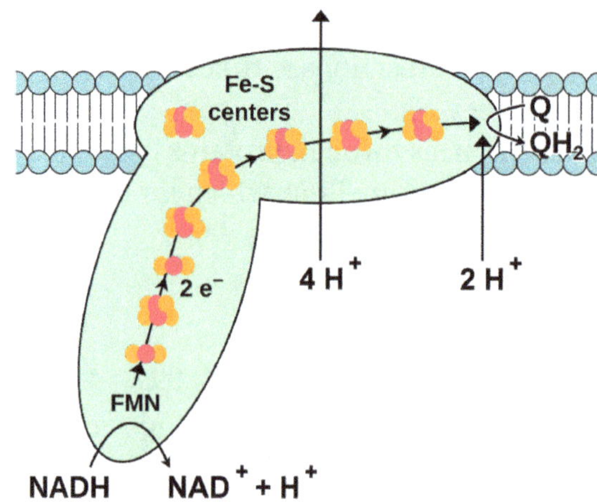

Complex I or NADH-Q oxidoreductase. The abbreviations are discussed in the text. In all diagrams of respiratory complexes in this article, the matrix is at the bottom, with the intermembrane space above.

NADH-coenzyme Q oxidoreductase, also known as *NADH dehydrogenase* or *complex I*, is the first protein in the electron transport chain. Complex I is a giant enzyme with the mammalian complex I having 46 subunits and a molecular mass of about 1,000 kilodaltons (kDa). The structure is known in detail only from a bacterium; in most organisms the complex resembles a boot with a large "ball" poking out from the membrane into the mitochondrion. The genes that encode the individual proteins are contained in both the cell nucleus and the mitochondrial genome, as is the case for many enzymes present in the mitochondrion.

The reaction that is catalyzed by this enzyme is the two electron oxidation of NADH by coenzyme Q10 or *ubiquinone* (represented as Q in the equation below), a lipid-soluble quinone that is found in the mitochondrion membrane:

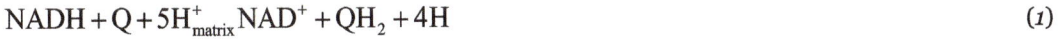

$$NADH + Q + 5H^{+}_{matrix} NAD^{+} + QH_2 + 4H \quad (1)$$

The start of the reaction, and indeed of the entire electron chain, is the binding of a NADH molecule to complex I and the donation of two electrons. The electrons enter complex I via a prosthetic group attached to the complex, flavin mononucleotide (FMN). The addition of electrons to FMN converts it to its reduced form, $FMNH_2$. The electrons are then transferred through a series of iron–sulfur clusters: the second kind of prosthetic group present in the complex. There are both [2Fe–2S] and [4Fe–4S] iron–sulfur clusters in complex I.

As the electrons pass through this complex, four protons are pumped from the matrix into the intermembrane space. Exactly how this occurs is unclear, but it seems to involve conformational changes in complex I that cause the protein to bind protons on the N-side of the membrane and release them on the P-side of the membrane. Finally, the electrons are transferred from the chain of iron–sulfur clusters to a ubiquinone molecule in the membrane. Reduction of ubiquinone also contributes to the generation of a proton gradient, as two protons are taken up from the matrix as it is reduced to ubiquinol (QH_2).

Succinate-Q Oxidoreductase (Complex II)

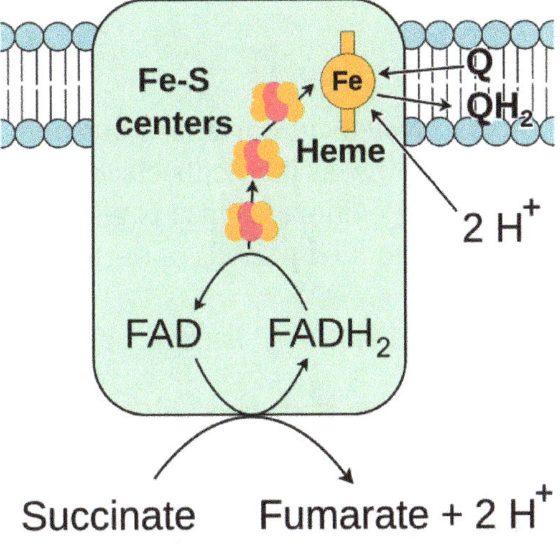

Complex II: Succinate-Q oxidoreductase.

Succinate-Q oxidoreductase, also known as *complex II* or *succinate dehydrogenase,* is a second entry point to the electron transport chain. It is unusual because it is the only enzyme that is part of both the citric acid cycle and the electron transport chain. Complex II consists of four protein subunits and contains a bound flavin adenine dinucleotide (FAD) cofactor, iron–sulfur clusters, and a heme group that does not participate in electron transfer to coenzyme Q, but is believed to be important in decreasing production of reactive oxygen species. It oxidizes succinate to fumarate and reduces ubiquinone. As this reaction releases less energy than the oxidation of NADH, complex II does not transport protons across the membrane and does not contribute to the proton gradient.

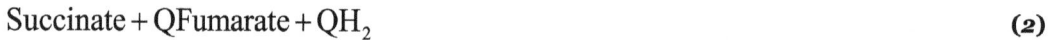

$$\text{Succinate} + \text{QFumarate} + \text{QH}_2 \quad (2)$$

In some eukaryotes, such as the parasitic worm *Ascaris suum*, an enzyme similar to complex II, fumarate reductase (menaquinol:fumarate oxidoreductase, or QFR), operates in reverse to oxidize ubiquinol and reduce fumarate. This allows the worm to survive in the anaerobic environment of the large intestine, carrying out anaerobic oxidative phosphorylation with fumarate as the electron acceptor. Another unconventional function of complex II is seen in the malaria parasite *Plasmodium falciparum*. Here, the reversed action of complex II as an oxidase is important in regenerating ubiquinol, which the parasite uses in an unusual form of pyrimidine biosynthesis.

Electron Transfer Flavoprotein-Q Oxidoreductase

Electron transfer flavoprotein-ubiquinone oxidoreductase (ETF-Q oxidoreductase), also known as *electron transferring-flavoprotein dehydrogenase,* is a third entry point to the electron transport chain. It is an enzyme that accepts electrons from electron-transferring flavoprotein in the mitochondrial matrix, and uses these electrons to reduce ubiquinone. This enzyme contains a flavin and a [4Fe–4S] cluster, but, unlike the other respiratory complexes, it attaches to the surface of the membrane and does not cross the lipid bilayer.

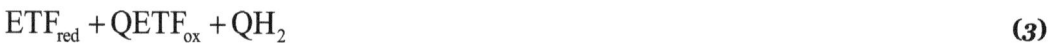

$$\text{ETF}_{\text{red}} + \text{QETF}_{\text{ox}} + \text{QH}_2 \quad (3)$$

In mammals, this metabolic pathway is important in beta oxidation of fatty acids and catabolism of amino acids and choline, as it accepts electrons from multiple acetyl-CoA dehydrogenases. In plants, ETF-Q oxidoreductase is also important in the metabolic responses that allow survival in extended periods of darkness.

Q-cytochrome C Oxidoreductase (Complex III)

Q-cytochrome c oxidoreductase is also known as *cytochrome c reductase, cytochrome* bc_1 *complex,* or simply *complex III*. In mammals, this enzyme is a dimer, with each subunit complex containing 11 protein subunits, an [2Fe-2S] iron–sulfur cluster and three cytochromes: one cytochrome c_1 and two b cytochromes. A cytochrome is a kind of electron-transferring protein that contains at least one heme group. The iron atoms inside complex III's heme groups alternate between a reduced ferrous (+2) and oxidized ferric (+3) state as the electrons are transferred through the protein.

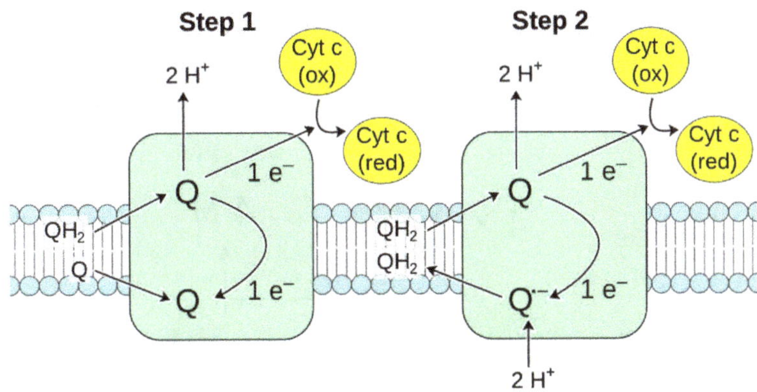

The two electron transfer steps in complex III: Q-cytochrome c oxidoreductase. After each step, Q (in the upper part of the figure) leaves the enzyme.

The reaction catalyzed by complex III is the oxidation of one molecule of ubiquinol and the reduction of two molecules of cytochrome c, a heme protein loosely associated with the mitochondrion. Unlike coenzyme Q, which carries two electrons, cytochrome c carries only one electron.

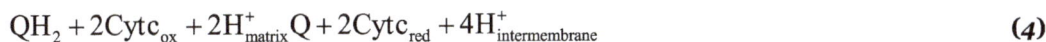

$$QH_2 + 2Cytc_{ox} + 2H^+_{matrix} Q + 2Cytc_{red} + 4H^+_{intermembrane} \quad (4)$$

As only one of the electrons can be transferred from the QH_2 donor to a cytochrome c acceptor at a time, the reaction mechanism of complex III is more elaborate than those of the other respiratory complexes, and occurs in two steps called the Q cycle. In the first step, the enzyme binds three substrates, first, QH_2, which is then oxidized, with one electron being passed to the second substrate, cytochrome c. The two protons released from QH_2 pass into the intermembrane space. The third substrate is Q, which accepts the second electron from the QH_2 and is reduced to $Q^{\cdot -}$, which is the ubisemiquinone free radical. The first two substrates are released, but this ubisemiquinone intermediate remains bound. In the second step, a second molecule of QH_2 is bound and again passes its first electron to a cytochrome c acceptor. The second electron is passed to the bound ubisemiquinone, reducing it to QH_2 as it gains two protons from the mitochondrial matrix. This QH_2 is then released from the enzyme.

As coenzyme Q is reduced to ubiquinol on the inner side of the membrane and oxidized to ubiquinone on the other, a net transfer of protons across the membrane occurs, adding to the proton gradient. The rather complex two-step mechanism by which this occurs is important, as it increases the efficiency of proton transfer. If, instead of the Q cycle, one molecule of QH_2 were used to directly reduce two molecules of cytochrome c, the efficiency would be halved, with only one proton transferred per cytochrome c reduced.

Cytochrome C Oxidase (Complex IV)

Cytochrome c oxidase, also known as *complex IV*, is the final protein complex in the electron transport chain. The mammalian enzyme has an extremely complicated structure and contains 13 subunits, two heme groups, as well as multiple metal ion cofactors – in all, three atoms of copper, one of magnesium and one of zinc.

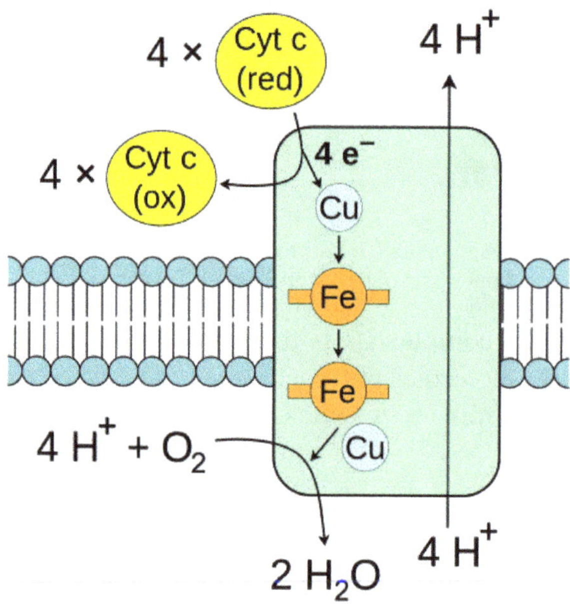

Complex IV: cytochrome c oxidase.

This enzyme mediates the final reaction in the electron transport chain and transfers electrons to oxygen, while pumping protons across the membrane. The final electron acceptor oxygen, which is also called the *terminal electron acceptor*, is reduced to water in this step. Both the direct pumping of protons and the consumption of matrix protons in the reduction of oxygen contribute to the proton gradient. The reaction catalyzed is the oxidation of cytochrome c and the reduction of oxygen:

$$4\mathrm{Cytc}_{\mathrm{red}} + \mathrm{O}_2 + 8\mathrm{H}^+_{\mathrm{matrix}} 4\mathrm{Cytc}_{\mathrm{ox}} + 2\mathrm{H}_2\mathrm{O} + 4\mathrm{H}^+_{\mathrm{intermembrane}} \quad (5)$$

Alternative Reductases and Oxidases

Many eukaryotic organisms have electron transport chains that differ from the much-studied mammalian enzymes described above. For example, plants have alternative NADH oxidases, which oxidize NADH in the cytosol rather than in the mitochondrial matrix, and pass these electrons to the ubiquinone pool. These enzymes do not transport protons, and, therefore, reduce ubiquinone without altering the electrochemical gradient across the inner membrane.

Another example of a divergent electron transport chain is the *alternative oxidase*, which is found in plants, as well as some fungi, protists, and possibly some animals. This enzyme transfers electrons directly from ubiquinol to oxygen.

The electron transport pathways produced by these alternative NADH and ubiquinone oxidases have lower ATP yields than the full pathway. The advantages produced by a shortened pathway are not entirely clear. However, the alternative oxidase is produced in response to stresses such as cold, reactive oxygen species, and infection by pathogens, as well as other factors that inhibit the full electron transport chain. Alternative pathways might, therefore, enhance an organisms' resistance to injury, by reducing oxidative stress.

Organization of Complexes

The original model for how the respiratory chain complexes are organized was that they diffuse freely and independently in the mitochondrial membrane. However, recent data suggest that the complexes might form higher-order structures called supercomplexes or "respirasomes." In this model, the various complexes exist as organized sets of interacting enzymes. These associations might allow channeling of substrates between the various enzyme complexes, increasing the rate and efficiency of electron transfer. Within such mammalian supercomplexes, some components would be present in higher amounts than others, with some data suggesting a ratio between complexes I/II/III/IV and the ATP synthase of approximately 1:1:3:7:4. However, the debate over this supercomplex hypothesis is not completely resolved, as some data do not appear to fit with this model.

Prokaryotic Electron Transport Chains

In contrast to the general similarity in structure and function of the electron transport chains in eukaryotes, bacteria and archaea possess a large variety of electron-transfer enzymes. These use an equally wide set of chemicals as substrates. In common with eukaryotes, prokaryotic electron transport uses the energy released from the oxidation of a substrate to pump ions across a membrane and generate an electrochemical gradient. In the bacteria, oxidative phosphorylation in *Escherichia coli* is understood in most detail, while archaeal systems are at present poorly understood.

The main difference between eukaryotic and prokaryotic oxidative phosphorylation is that bacteria and archaea use many different substances to donate or accept electrons. This allows prokaryotes to grow under a wide variety of environmental conditions. In *E. coli*, for example, oxidative phosphorylation can be driven by a large number of pairs of reducing agents and oxidizing agents, which are listed below. The midpoint potential of a chemical measures how much energy is released when it is oxidized or reduced, with reducing agents having negative potentials and oxidizing agents positive potentials.

Respiratory enzymes and substrates in *E. coli*.		
Respiratory enzyme	**Redox pair**	**Midpoint potential (Volts)**
Formate dehydrogenase	Bicarbonate / Formate	−0.43
Hydrogenase	Proton / Hydrogen	−0.42
NADH dehydrogenase	NAD^+ / NADH	−0.32
Glycerol-3-phosphate dehydrogenase	DHAP / Gly-3-P	−0.19
Pyruvate oxidase	Acetate + Carbon dioxide / Pyruvate	?
Lactate dehydrogenase	Pyruvate / Lactate	−0.19
D-amino acid dehydrogenase	2-oxoacid + ammonia / D-amino acid	?
Glucose dehydrogenase	Gluconate / Glucose	−0.14
Succinate dehydrogenase	Fumarate / Succinate	+0.03
Ubiquinol oxidase	Oxygen / Water	+0.82
Nitrate reductase	Nitrate / Nitrite	+0.42
Nitrite reductase	Nitrite / Ammonia	+0.36
Dimethyl sulfoxide reductase	DMSO / DMS	+0.16
Trimethylamine *N*-oxide reductase	TMAO / TMA	+0.13
Fumarate reductase	Fumarate / Succinate	+0.03

As shown above, *E. coli* can grow with reducing agents such as formate, hydrogen, or lactate as electron donors, and nitrate, DMSO, or oxygen as acceptors. The larger the difference in midpoint potential between an oxidizing and reducing agent, the more energy is released when they react. Out of these compounds, the succinate/fumarate pair is unusual, as its midpoint potential is close to zero. Succinate can therefore be oxidized to fumarate if a strong oxidizing agent such as oxygen is available, or fumarate can be reduced to succinate using a strong reducing agent such as formate. These alternative reactions are catalyzed by succinate dehydrogenase and fumarate reductase, respectively.

Some prokaryotes use redox pairs that have only a small difference in midpoint potential. For example, nitrifying bacteria such as *Nitrobacter* oxidize nitrite to nitrate, donating the electrons to oxygen. The small amount of energy released in this reaction is enough to pump protons and generate ATP, but not enough to produce NADH or NADPH directly for use in anabolism. This problem is solved by using a nitrite oxidoreductase to produce enough proton-motive force to run part of the electron transport chain in reverse, causing complex I to generate NADH.

Prokaryotes control their use of these electron donors and acceptors by varying which enzymes are produced, in response to environmental conditions. This flexibility is possible because different oxidases and reductases use the same ubiquinone pool. This allows many combinations of enzymes to function together, linked by the common ubiquinol intermediate. These respiratory chains therefore have a modular design, with easily interchangeable sets of enzyme systems.

In addition to this metabolic diversity, prokaryotes also possess a range of isozymes – different enzymes that catalyze the same reaction. For example, in *E. coli*, there are two different types of ubiquinol oxidase using oxygen as an electron acceptor. Under highly aerobic conditions, the cell uses an oxidase with a low affinity for oxygen that can transport two protons per electron. However, if levels of oxygen fall, they switch to an oxidase that transfers only one proton per electron, but has a high affinity for oxygen.

ATP Synthase (Complex V)

ATP synthase, also called *complex V*, is the final enzyme in the oxidative phosphorylation pathway. This enzyme is found in all forms of life and functions in the same way in both prokaryotes and eukaryotes. The enzyme uses the energy stored in a proton gradient across a membrane to drive the synthesis of ATP from ADP and phosphate (P_i). Estimates of the number of protons required to synthesize one ATP have ranged from three to four, with some suggesting cells can vary this ratio, to suit different conditions.

$$ADP + P_i + 4H+_{intermembrane} \rightleftharpoons ATP + H_2O + 4H+_{matrix} \quad (7)$$

This phosphorylation reaction is an equilibrium, which can be shifted by altering the proton-motive force. In the absence of a proton-motive force, the ATP synthase reaction will run from right to left, hydrolyzing ATP and pumping protons out of the matrix across the membrane. However, when the proton-motive force is high, the reaction is forced to run in the opposite direction; it proceeds from left to right, allowing protons to flow down their concentration gradient and turning ADP into ATP. Indeed, in the closely related vacuolar type H+-ATPases, the hydrolysis reaction is used to acidify cellular compartments, by pumping protons and hydrolysing ATP.

ATP synthase is a massive protein complex with a mushroom-like shape. The mammalian enzyme complex contains 16 subunits and has a mass of approximately 600 kilodaltons. The portion embedded within the membrane is called F_O and contains a ring of c subunits and the proton channel. The stalk and the ball-shaped headpiece is called F_1 and is the site of ATP synthesis. The ball-shaped complex at the end of the F_1 portion contains six proteins of two different kinds (three α subunits and three β subunits), whereas the "stalk" consists of one protein: the γ subunit, with the tip of the stalk extending into the ball of α and β subunits. Both the α and β subunits bind nucleotides, but only the β subunits catalyze the ATP synthesis reaction. Reaching along the side of the F_1 portion and back into the membrane is a long rod-like subunit that anchors the α and β subunits into the base of the enzyme.

As protons cross the membrane through the channel in the base of ATP synthase, the F_O proton-driven motor rotates. Rotation might be caused by changes in the ionization of amino acids in the ring of c subunits causing electrostatic interactions that propel the ring of c subunits past the proton channel. This rotating ring in turn drives the rotation of the central axle (the γ subunit stalk) within the α and β subunits. The α and β subunits are prevented from rotating themselves by the side-arm, which acts as a stator. This movement of the tip of the γ subunit within the ball of α and β subunits provides the energy for the active sites in the β subunits to undergo a cycle of movements that produces and then releases ATP.

This ATP synthesis reaction is called the *binding change mechanism* and involves the active site of a β subunit cycling between three states. In the "open" state, ADP and phosphate enter the active site (shown in brown in the diagram). The protein then closes up around the molecules and binds them loosely – the "loose" state (shown in red). The enzyme then changes shape again and forces these molecules together, with the active site in the resulting "tight" state (shown in pink) binding the newly produced ATP molecule with very high affinity. Finally, the active site cycles back to the open state, releasing ATP and binding more ADP and phosphate, ready for the next cycle.

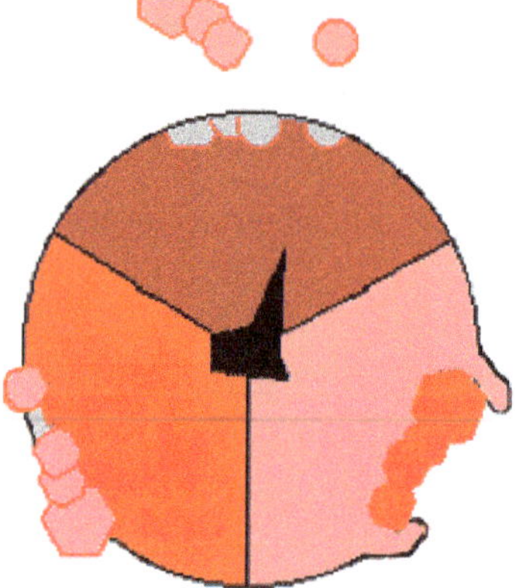

Mechanism of ATP synthase. ATP is shown in red, ADP and phosphate in pink and the rotating γ subunit in black.

In some bacteria and archaea, ATP synthesis is driven by the movement of sodium ions through the cell membrane, rather than the movement of protons. Archaea such as *Methanococcus* also contain the A_1A_0 synthase, a form of the enzyme that contains additional proteins with little similarity in sequence to other bacterial and eukaryotic ATP synthase subunits. It is possible that, in some species, the A_1A_0 form of the enzyme is a specialized sodium-driven ATP synthase, but this might not be true in all cases.

Reactive Oxygen Species

Molecular oxygen is an ideal terminal electron acceptor because it is a strong oxidizing agent. The reduction of oxygen does involve potentially harmful intermediates. Although the transfer of four electrons and four protons reduces oxygen to water, which is harmless, transfer of one or two electrons produces superoxide or peroxide anions, which are dangerously reactive.

$$O_2 \xrightarrow{e^-} \underset{\text{Superoxide}}{O_2^{\bullet-}} \xrightarrow{e^-} \underset{\text{Peroxide}}{O_2^{2-}} \qquad (8)$$

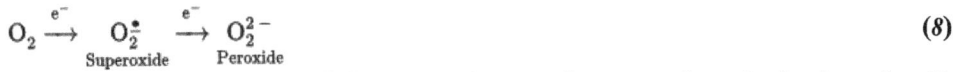

These reactive oxygen species and their reaction products, such as the hydroxyl radical, are very harmful to cells, as they oxidize proteins and cause mutations in DNA. This cellular damage might contribute to disease and is proposed as one cause of aging.

The cytochrome c oxidase complex is highly efficient at reducing oxygen to water, and it releases very few partly reduced intermediates; however small amounts of superoxide anion and peroxide are produced by the electron transport chain. Particularly important is the reduction of coenzyme Q in complex III, as a highly reactive ubisemiquinone free radical is formed as an intermediate in the Q cycle. This unstable species can lead to electron "leakage" when electrons transfer directly to oxygen, forming superoxide. As the production of reactive oxygen species by these proton-pumping complexes is greatest at high membrane potentials, it has been proposed that mitochondria regulate their activity to maintain the membrane potential within a narrow range that balances ATP production against oxidant generation. For instance, oxidants can activate uncoupling proteins that reduce membrane potential.

To counteract these reactive oxygen species, cells contain numerous antioxidant systems, including antioxidant vitamins such as vitamin C and vitamin E, and antioxidant enzymes such as superoxide dismutase, catalase, and peroxidases, which detoxify the reactive species, limiting damage to the cell.

Inhibitors

There are several well-known drugs and toxins that inhibit oxidative phosphorylation. Although any one of these toxins inhibits only one enzyme in the electron transport chain, inhibition of any step in this process will halt the rest of the process. For example, if oligomycin inhibits ATP synthase, protons cannot pass back into the mitochondrion. As a result, the proton pumps are unable to operate, as the gradient becomes too strong for them to overcome. NADH is then no longer oxidized and the citric acid cycle ceases to operate because the concentration of NAD^+ falls below the concentration that these enzymes can use.

Compounds	**Use**	**Site of action**	**Effect on oxidative phosphorylation**
Cyanide Carbon monoxide Azide Hydrogen sulfide	Poisons	Complex IV	Inhibit the electron transport chain by binding more strongly than oxygen to the Fe–Cu center in cytochrome c oxidase, preventing the reduction of oxygen.
Oligomycin	Antibiotic	Complex V	Inhibits ATP synthase by blocking the flow of protons through the F_o subunit.

CCCP 2,4-Dinitrophenol	Poisons, weight-loss	Inner membrane	Ionophores that disrupt the proton gradient by carrying protons across a membrane. This ionophore uncouples proton pumping from ATP synthesis because it carries protons across the inner mitochondrial membrane.
Rotenone	Pesticide	Complex I	Prevents the transfer of electrons from complex I to ubiquinone by blocking the ubiquinone-binding site.
Malonate and oxaloacetate	Poisons	Complex II	Competitive inhibitors of succinate dehydrogenase (complex II).
Antimycin A	Piscicide	Complex III	Binds to the Qi site of cytochrome c reductase, thereby inhibiting the oxidation of ubiquinol.

Not all inhibitors of oxidative phosphorylation are toxins. In brown adipose tissue, regulated proton channels called uncoupling proteins can uncouple respiration from ATP synthesis. This rapid respiration produces heat, and is particularly important as a way of maintaining body temperature for hibernating animals, although these proteins may also have a more general function in cells' responses to stress.

History

The field of oxidative phosphorylation began with the report in 1906 by Arthur Harden of a vital role for phosphate in cellular fermentation, but initially only sugar phosphates were known to be involved. However, in the early 1940s, the link between the oxidation of sugars and the generation of ATP was firmly established by Herman Kalckar, confirming the central role of ATP in energy transfer that had been proposed by Fritz Albert Lipmann in 1941. Later, in 1949, Morris Friedkin and Albert L. Lehninger proved that the coenzyme NADH linked metabolic pathways such as the citric acid cycle and the synthesis of ATP. The term *oxidative phosphorylation* was coined by Volodymyr Belitser (uk) in 1939.

For another twenty years, the mechanism by which ATP is generated remained mysterious, with scientists searching for an elusive "high-energy intermediate" that would link oxidation and phosphorylation reactions. This puzzle was solved by Peter D. Mitchell with the publication of the chemiosmotic theory in 1961. At first, this proposal was highly controversial, but it was slowly accepted and Mitchell was awarded a Nobel prize in 1978. Subsequent research concentrated on purifying and characterizing the enzymes involved, with major contributions being made by David E. Green on the complexes of the electron-transport chain, as well as Efraim Racker on the ATP synthase. A critical step towards solving the mechanism of the ATP synthase was provided by Paul D. Boyer, by his development in 1973 of the "binding change" mechanism, followed by his radical proposal of rotational catalysis in 1982. More recent work has included structural studies on the enzymes involved in oxidative phosphorylation by John E. Walker, with Walker and Boyer being awarded a Nobel Prize in 1997.

Adenosine Triphosphate

Adenosine triphosphate (ATP) is a nucleoside triphosphate, a small molecule used in cells as a coenzyme. It is often referred to as the "molecular unit of currency" of intracellular energy transfer.

ATP transports chemical energy within cells for metabolism. Most cellular functions need energy to be carried out: synthesis of proteins, synthesis of membranes, movement of the cell, cellular division, etc. need energy in order to be made. The ATP is the molecule that carries energy to the place where the energy is needed. When ATP breaks into ADP (Adenosine diphosphate) and P_i (phosphate), the breakdown of the last covalent link of phosphate (a simple -Po4) liberates energy that is used in reactions where it is needed.

It is one of the end products of photophosphorylation, aerobic respiration, and fermentation, and is used by enzymes and structural proteins in many cellular processes, including biosynthetic reactions, motility, and cell division. One molecule of ATP contains adenosine, ribose, and three phosphate groups, and it is produced by a wide variety of enzymes, including ATP synthase, from adenosine diphosphate (ADP) or adenosine monophosphate (AMP) and various phosphate group donors. Substrate-level phosphorylation, oxidative phosphorylation in cellular respiration, and photophosphorylation in photosynthesis are three major mechanisms of ATP biosynthesis.

Metabolic processes that use ATP as an energy source convert it back into its precursors. ATP is therefore continuously recycled in organisms: the human body, which on average contains only 250 grams (8.8 oz) of ATP, turns over its own body weight equivalent in ATP each day.

ATP is used as a substrate in signal transduction pathways by kinases that phosphorylate proteins and lipids. It is also used by adenylate cyclase, which uses ATP to produce the second messenger molecule cyclic AMP. The ratio between ATP and AMP is used as a way for a cell to sense how much energy is available and control the metabolic pathways that produce and consume ATP. Apart from its roles in signaling and energy metabolism, ATP is also incorporated into nucleic acids by polymerases in the process of transcription. ATP is the neurotransmitter believed to signal the sense of taste.

The structure of this molecule consists of a purine base (adenine) attached by the 9′ nitrogen atom to the 1′ carbon atom of a pentose sugar (ribose). Three phosphate groups are attached at the 5′ carbon atom of the pentose sugar. It is the addition and removal of these phosphate groups that inter-convert ATP, ADP and AMP. When ATP is used in DNA synthesis, the ribose sugar is first converted to deoxyribose by ribonucleotide reductase.

ATP was discovered in 1929 by Karl Lohmann, and independently by Cyrus Fiske and Yellapragada Subbarow of Harvard Medical School, but its correct structure was not determined until some years later. It was proposed to be the intermediary molecule between energy-yielding and energy-requiring reactions in cells by Fritz Albert Lipmann in 1941. It was first artificially synthesized by Alexander Todd in 1948.

Physical and Chemical Properties

ATP consists of adenosine – composed of an adenine ring and a ribose sugar – and three phosphate groups (triphosphate). The phosphoryl groups, starting with the group closest to the ribose, are referred to as the alpha (α), beta (β), and gamma (γ) phosphates. Consequently, it is closely related to the adenosine nucleotide, a monomer of RNA. ATP is highly soluble in water and is quite stable in solutions between pH 6.8 and 7.4, but is rapidly hydrolysed at extreme pH. Consequently, ATP is best stored as an anhydrous salt.

ATP is an unstable molecule in unbuffered water, in which it hydrolyses to ADP and phosphate. This is because the strength of the bonds between the phosphate groups in ATP is less than the strength of the hydrogen bonds (hydration bonds), between its products (ADP and phosphate), and water. Thus, if ATP and ADP are in chemical equilibrium in water, almost all of the ATP will eventually be converted to ADP. A system that is far from equilibrium contains Gibbs free energy, and is capable of doing work. Living cells maintain the ratio of ATP to ADP at a point ten orders of magnitude from equilibrium, with ATP concentrations fivefold higher than the concentration of ADP. This displacement from equilibrium means that the hydrolysis of ATP in the cell releases a large amount of free energy.

Two phosphoanhydride bonds (those that connect adjacent phosphates) in an ATP molecule are responsible for the high energy content of this molecule. In the context of biochemical reactions, these anhydride bonds are frequently – and sometimes controversially – referred to as *high-energy bonds* (despite the fact it takes energy to break bonds). Energy stored in ATP may be released upon hydrolysis of the anhydride bonds. The primary phosphate group on the ATP molecule that is hydrolyzed when energy is needed to drive anabolic reactions is the γ-phosphate group. Located the farthest from the ribose sugar, it has a higher energy of hydrolysis than either the α- or β-phosphate. The bonds formed after hydrolysis – or the phosphorylation of a residue by ATP – are lower in energy than the phosphoanhydride bonds of ATP. During enzyme-catalyzed hydrolysis of ATP or phosphorylation by ATP, the available free energy can be harnessed by a living system to do work.

Any unstable system of potentially reactive molecules could potentially serve as a way of storing free energy, if the cell maintained their concentration far from the equilibrium point of the reaction. However, as is the case with most polymeric biomolecules, the breakdown of RNA, DNA, and ATP into simpler monomers is driven by both energy-release and entropy-increase considerations, in both standard concentrations, and also those concentrations encountered within the cell.

The standard amount of energy released from hydrolysis of ATP can be calculated from the changes in energy under non-natural (standard) conditions, then correcting to biological concentrations. The net change in heat energy (enthalpy) at standard temperature and pressure of the decomposition of ATP into hydrated ADP and hydrated inorganic phosphate is −30.5 kJ/mol, with a change in free energy of 3.4 kJ/mol. The energy released by cleaving either a phosphate (P_i) or pyrophosphate (PP_i) unit from ATP at standard state of 1 M are:

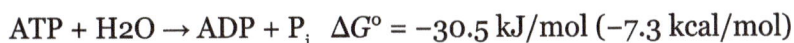

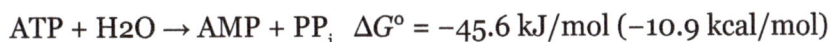

$$ATP + H2O \rightarrow ADP + P_i \quad \Delta G^o = -30.5 \text{ kJ/mol } (-7.3 \text{ kcal/mol})$$

$$ATP + H2O \rightarrow AMP + PP_i \quad \Delta G^o = -45.6 \text{ kJ/mol } (-10.9 \text{ kcal/mol})$$

These values can be used to calculate the change in energy under physiological conditions and the cellular ATP/ADP ratio. However, a more representative value (which takes AMP into consideration) called the Energy charge is increasingly being employed. The values given for the Gibbs free energy for this reaction are dependent on a number of factors, including overall ionic strength and the presence of alkaline earth metal ions such as Mg2+and Ca2+. Under typical cellular conditions, ΔG is approximately −57 kJ/mol (−14 kcal/mol).

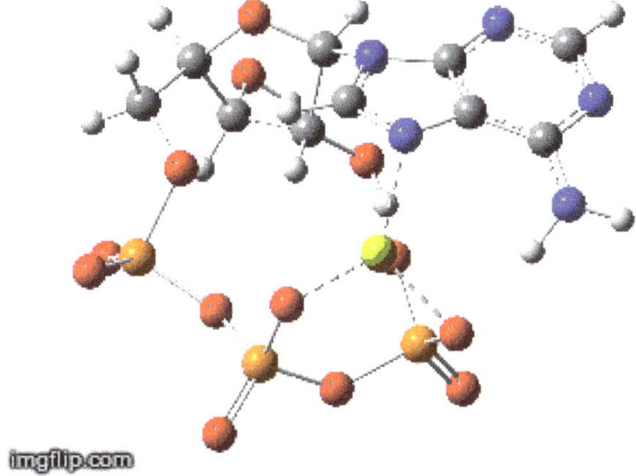

This image shows a full 360-degree rotation of a single, gas-phase magnesium-ATP chelate with a charge of −2. The molecule was optimized at the UB3LYP/6-311++G(d,p) theoretical level and the atomic connectivity modified by the human optimizer to reflect the probable electronic structure.

Ionization in Biological Systems

ATP (adenosine triphosphate) has multiple groups with different acid dissociation constants. In neutral solution, ionized ATP exists mostly as ATP^{4-}, with a small proportion of ATP^{3-}. As ATP has several negatively charged groups in neutral solution, it can chelate metals with very high affinity. The binding constant for various metal ions are (given as per mole) as Mg2+(9554), Na+(13), Ca2+(3722), K+(8), Sr2+(1381) and Li+(25). Due to the strength of these interactions, ATP exists in the cell mostly in a complex with Mg2+.

Biosynthesis

The ATP concentration inside the cell is typically 1–10 mM. ATP can be produced by redox reactions using simple and complex sugars (carbohydrates) or lipids as an energy source. For complex fuels to be synthesized into ATP, they first need to be broken down into smaller, more simple molecules. Carbohydrates are hydrolysed into simple sugars, such as glucose and fructose. Fats (triglycerides) are metabolised to give fatty acids and glycerol.

The overall process of oxidizing glucose to carbon dioxide is known as cellular respiration and can produce about 30 molecules of ATP from a single molecule of glucose. ATP can be produced by a number of distinct cellular processes; the three main pathways used to generate energy in eukaryotic organisms are glycolysis and the citric acid cycle/oxidative phosphorylation, both components of cellular respiration; and beta-oxidation. The majority of this ATP production by a non-photosynthetic aerobic eukaryote takes place in the mitochondria, which can make up nearly 25% of the total volume of a typical cell.

Glycolysis

In glycolysis, glucose and glycerol are metabolized to pyruvate via the glycolytic pathway. In most organisms, this process occurs in the cytosol, but, in some protozoa such as the kinetoplastids, this is carried out in a specialized organelle called the glycosome. Glycolysis generates a net two molecules of ATP through substrate phosphorylation catalyzed by two enzymes: PGK and pyruvate kinase. Two molecules of NADH are also produced, which can be oxidized via the electron transport chain and result in the generation of additional ATP by ATP synthase. The pyruvate generated as an end-product of glycolysis is a substrate for the Krebs Cycle.

Glucose

In the mitochondrion, pyruvate is oxidized by the pyruvate dehydrogenase complex to the acetyl group, which is fully oxidized to carbon dioxide by the citric acid cycle (also known as the Krebs cycle). Every "turn" of the citric acid cycle produces two molecules of carbon dioxide, one molecule of the ATP equivalent guanosine triphosphate (GTP) through substrate-level phosphorylation catalyzed by succinyl-CoA synthetase, three molecules of the reduced coenzyme NADH, and one molecule of the reduced coenzyme $FADH_2$. Both of these latter molecules are recycled to their oxidized states (NAD^+ and FAD, respectively) via the electron transport chain, which generates additional ATP by oxidative phosphorylation. The oxidation of an NADH molecule results in the synthesis of 2–3 ATP molecules, and the oxidation of one $FADH_2$ yields between 1–2 ATP molecules. The majority of cellular ATP is generated by this process. Although the citric acid cycle itself does not involve molecular oxygen, it is an obligately aerobic process because O_2 is needed to recycle the reduced NADH and $FADH_2$ to their oxidized states. In the absence of oxygen the citric acid cycle will cease to function due to the lack of available NAD^+ and FAD.

The generation of ATP by the mitochondrion from cytosolic NADH relies on the malate-aspartate shuttle (and to a lesser extent, the glycerol-phosphate shuttle) because the inner mitochondrial membrane is impermeable to NADH and NAD^+. Instead of transferring the generated NADH, a malate dehydrogenase enzyme converts oxaloacetate to malate, which is translocated to the mitochondrial matrix. Another malate dehydrogenase-catalyzed reaction occurs in the opposite direction, producing oxaloacetate and NADH from the newly transported malate and the mitochondrion's interior store of NAD^+. A transaminase converts the oxaloacetate to aspartate for transport back across the membrane and into the intermembrane space.

In oxidative phosphorylation, the passage of electrons from NADH and $FADH_2$ through the electron transport chain powers the pumping of protons out of the mitochondrial matrix and into the intermembrane space. This creates a proton motive force that is the net effect of a pH gradient and an electric potential gradient across the inner mitochondrial membrane. Flow of protons down this potential gradient – that is, from the intermembrane space to the matrix – provides the driving force for ATP synthesis by ATP synthase. This enzyme contains a rotor subunit that physically rotates relative to the static portions of the protein during ATP synthesis.

Most of the ATP synthesized in the mitochondria will be used for cellular processes in the cytosol; thus it must be exported from its site of synthesis in the mitochondrial matrix. The inner membrane contains an antiporter, the ADP/ATP translocase, which is an integral membrane protein used to exchange newly synthesized ATP in the matrix for ADP in the intermembrane space. This translocase is driven by the membrane potential, as it results in the movement of about 4 negative charges out of the mitochondrial membrane in exchange for 3 negative charges moved inside. However, it is also necessary to transport phosphate into the mitochondrion; the phosphate carrier moves a proton in with each phosphate, partially dissipating the proton gradient.

Beta Oxidation

Fatty acids can also be broken down to acetyl-CoA by beta-oxidation. Each round of this cycle reduces the length of the acyl chain by two carbon atoms and produces one NADH and one $FADH_2$ molecule, which are used to generate ATP by oxidative phosphorylation. Because NADH and $FADH_2$ are energy-rich molecules, dozens of ATP molecules can be generated by the beta-oxidation of a single long acyl chain. The high energy yield of this process and the compact storage of fat explain why it is the most dense source of dietary calories.

Fermentation

Fermentation entails the generation of energy via the process of substrate-level phosphorylation in the absence of a respiratory electron transport chain. In most eukaryotes, glucose is used as both an energy store and an electron donor. The equation for the oxidation of glucose to lactic acid is:

$$C6H12O6 \rightarrow 2\ CH3CH(OH)COOH + 2\ ATP$$

Anaerobic Respiration

Anaerobic respiration is the process of respiration using an electron acceptor other than O 2. In prokaryotes, multiple electron acceptors can be used in anaerobic respiration. These include nitrate, sulfate or carbon dioxide. These processes lead to the ecologically important processes of denitrification, sulfate reduction and acetogenesis, respectively.

ATP Replenishment by Nucleoside Diphosphate Kinases

ATP can also be synthesized through several so-called "replenishment" reactions catalyzed by the enzyme families of nucleoside diphosphate kinases (NDKs), which use other nucleoside triphosphates as a high-energy phosphate donor, and the ATP:guanido-phosphotransferase family.

ATP Production During Photosynthesis

In plants, ATP is synthesized in thylakoid membrane of the chloroplast during the light-dependent reactions of photosynthesis in a process called photophosphorylation. Here, light energy is used to pump protons across the chloroplast membrane. This produces a proton-motive force and this drives the ATP synthase, exactly as in oxidative phosphorylation. Some of the ATP produced in the chloroplasts is consumed in the Calvin cycle, which produces triose sugars.

ATP Recycling

The total quantity of ATP in the human body is about 0.2 moles. The majority of ATP is not usually synthesised *de novo*, but is generated from ADP by the aforementioned processes. Thus, at any given time, the total amount of ATP + ADP remains fairly constant.

The energy used by human cells requires the hydrolysis of 100 to 150 moles of ATP daily, which is around 50 to 75 kg. A human will typically use up his or her body weight of ATP over the course of the day. This means that each ATP molecule is recycled 500 to 750 times during a single day (100 / 0.2 = 500). ATP cannot be stored, hence its consumption closely follows its synthesis. However a total of around 5 g of ATP is used by cell processes at any time in the body.

Regulation of Biosynthesis

ATP production in an aerobic eukaryotic cell is tightly regulated by allosteric mechanisms, by feedback effects, and by the substrate concentration dependence of individual enzymes within the glycolysis and oxidative phosphorylation pathways. Key control points occur in enzymatic reactions that are so energetically favorable that they are effectively irreversible under physiological conditions.

In glycolysis, hexokinase is directly inhibited by its product, glucose-6-phosphate, and pyruvate kinase is inhibited by ATP itself. The main control point for the glycolytic pathway is phosphofructokinase (PFK), which is allosterically inhibited by high concentrations of ATP and activated by high concentrations of AMP. The inhibition of PFK by ATP is unusual, since ATP is also a substrate in the reaction catalyzed by PFK; the biologically active form of the enzyme is a tetramer that exists in two possible conformations, only one of which binds the second substrate fructose-6-phosphate (F6P). The protein has two binding sites for ATP – the active site is accessible in either protein conformation, but ATP binding to the inhibitor site stabilizes the conformation that binds F6P poorly. A number of other small molecules can compensate for the ATP-induced shift in equilibrium conformation and reactivate PFK, including cyclic AMP, ammonium ions, inorganic phosphate, and fructose-1,6- and -2,6-biphosphate.

The citric acid cycle is regulated mainly by the availability of key substrates, particularly the ratio of NAD^+ to NADH and the concentrations of calcium, inorganic phosphate, ATP, ADP, and AMP. Citrate – the molecule that gives its name to the cycle – is a feedback inhibitor of citrate synthase and also inhibits PFK, providing a direct link between the regulation of the citric acid cycle and glycolysis.

In oxidative phosphorylation, the key control point is the reaction catalyzed by cytochrome c oxidase, which is regulated by the availability of its substrate – the reduced form of cytochrome c. The amount of reduced cytochrome c available is directly related to the amounts of other substrates:

$$\tfrac{1}{2}\,\text{NADH} + \text{cyt c}_{\text{ox}} + \text{ADP} + \text{P}_{\text{i}} \rightleftharpoons \tfrac{1}{2}\,\text{NAD}^+ + \text{cyt c}_{\text{red}} + \text{ATP}$$

which directly implies this equation:

$$\frac{[\text{cyt c}_{\text{red}}]}{[\text{cyt c}_{\text{ox}}]} = \left(\frac{[\text{NADH}]}{[\text{NAD}]^+}\right)^{\frac{1}{2}} \left(\frac{[\text{ADP}][\text{P}_{\text{i}}]}{[\text{ATP}]}\right) K_{eq}$$

Thus, a high ratio of [NADH] to [NAD^+] or a high ratio of [ADP][P_i] to [ATP] imply a high amount of reduced cytochrome c and a high level of cytochrome c oxidase activity. An additional level of regulation is introduced by the transport rates of ATP and NADH between the mitochondrial matrix and the cytoplasm.

Functions in Cells

Metabolism, Synthesis, and Active Transport

ATP is consumed in the cell by energy-requiring (endergonic) processes and can be generated by energy-releasing (exergonic) processes. In this way ATP transfers energy between spatially separate metabolic reactions. ATP is the main energy source for the majority of cellular functions. This includes the synthesis of macromolecules, including DNA and RNA, and proteins. ATP also plays a critical role in the transport of macromolecules across cell membranes, e.g. exocytosis and endocytosis.

Roles in Cell Structure and Locomotion

ATP is critically involved in maintaining cell structure by facilitating assembly and disassembly of elements of the cytoskeleton. In a related process, ATP is required for the shortening of actin and myosin filament crossbridges required for muscle contraction. This latter process is one of the main energy requirements of animals and is essential for locomotion and respiration.

Cell Signalling

Extracellular Signalling

Extracellular ATP (eATP) is also a signalling molecule. ATP, ADP, or adenosine are recognised by purinergic receptors, or purinoreceptors, which might be the most abundant receptors in mammalian tissues.

In humans, this signalling role is important in both the central and peripheral nervous system. Activity-dependent release of ATP from synapses, axons and glia activates purinergic membrane receptors known as P2. The *P2Y* receptors are G protein-coupled receptors, which are *metabotropic*, and primarily modulate intracellular calcium and, to a lesser extent, cyclic AMP levels. Though named between $P2Y_1$ and $P2Y_{15}$, only nine members of the P2Y family have been cloned, and some are only related through weak homology and several ($P2Y_5$, $P2Y_7$, $P2Y_9$, $P2Y_{10}$) do not function as receptors that raise cytosolic calcium. The *P2X ionotropic* receptor subgroup comprises seven members ($P2X_1$–$P2X_7$), which are ligand-gated Ca2+-permeable ion channels that open when bound to an extracellular purine nucleotide, like ATP. In contrast to P2 receptors (agonist rank order of potency: ATP > ADP > AMP > ADO), purinergic nucleoside triphosphates like ATP are not strong agonists of P1 receptors, which are strongly activated by adenosine and other nucleosides (ADO > AMP > ADP > ATP). P1 receptors have A1, A2a, A2b, and A3 subtypes (the "A" is standard nomenclature for indicating an *adenosine receptor* subtype), all of which are G protein-coupled receptors, A1 and A3 being coupled to Gi, and A2a and A2b being coupled to Gs. All adenosine receptors were shown to activate at least one subfamily of mitogen-activated protein kinases. The actions of adenosine are often antagonistic or synergistic to the actions of ATP. In the CNS, adenosine has multiple functions, such as modulation of neural development, neuron and glial signalling and the control of innate and adaptive immune systems.

Intracellular Signaling

ATP is critical in signal transduction processes. It is used by kinases as the source of phosphate groups in their phosphate transfer reactions. Kinase activity on substrates such as proteins or membrane lipids are a common form of signal transduction. Phosphorylation of a protein by a kinase can activate this cascade such as the mitogen-activated protein kinase cascade.

ATP is also used by adenylate cyclase and is transformed to the second messenger molecule cyclic AMP, which is involved in triggering calcium signals by the release of calcium from intracellular stores. This form of signal transduction is particularly important in brain function, although it is involved in the regulation of a multitude of other cellular processes.

DNA and RNA Synthesis

In all known organisms, the Deoxyribonucleotides that make up DNA are synthesized by the action of ribonucleotide reductase (RNR) enzymes on their corresponding ribonucleotides. These enzymes reduce the sugar residue from ribose to deoxyribose by removing oxygen from the 2′ hydroxyl group; the substrates are ribonucleoside diphosphates and the products deoxyribonucleoside diphosphates (the latter are denoted dADP, dCDP, dGDP, and dUDP respectively.) All ribonucleotide reductase enzymes use a common sulfhydryl radical mechanism reliant on reactive cysteine residues that oxidize to form disulfide bonds in the course of the reaction. RNR enzymes are recycled by reaction with thioredoxin or glutaredoxin.

The regulation of RNR and related enzymes maintains a balance of dNTPs relative to each other and relative to NTPs in the cell. Very low dNTP concentration inhibits DNA synthesis and DNA repair and is lethal to the cell, while an abnormal ratio of dNTPs is mutagenic due to the increased likelihood of the DNA polymerase incorporating the wrong dNTP during DNA synthesis. Regulation of or differential specificity of RNR has been proposed as a mechanism for alterations in the relative sizes of intracellular dNTP pools under cellular stress such as hypoxia.

In the synthesis of the nucleic acid RNA, adenosine derived from ATP is one of the four nucleotides incorporated directly into RNA molecules by RNA polymerases. The energy driving this polymerization comes from cleaving off a pyrophosphate (two phosphate groups). The process is similar in DNA biosynthesis, except that ATP is reduced to the deoxyribonucleotide dATP, before incorporation into DNA.

Amino Acid Activation in Protein Synthesis

Aminoacyl-tRNA synthetase enzymes utilize ATP as an energy source to attach a tRNA molecule to its specific amino acid, forming an aminoacyl-tRNA complex, ready for translation at ribosomes. The energy is made available by ATP hydrolysis to adenosine monophosphate (AMP) as two phosphate groups are removed. Amino acid activation refers to the attachment of an amino acid to its Transfer RNA (tRNA). Aminoacyl transferase binds Adenosine triphosphate (ATP) to amino acid, PP is released. Aminoacyl transferase binds AMP-amino acid to tRNA. The AMP is used in this step.

Amino Acid Activation

During amino acid activation the amino acids (aa) are attached to their corresponding tRNA. The coupling reactions are catalysed by a group of enzymes called aminoacyl-tRNA synthetases (named after the reaction product aminoacyl-tRNA or aa-tRNA). The coupling reaction proceeds in two steps:

1. aa + ATP aa-AMP + PP_i
2. aa-AMP + tRNA aa-tRNA + AMP

The amino acid is coupled to the penultimate nucleotide at the 3′-end of the tRNA (the A in the sequence CCA) via an ester bond (roll over in illustration). The formation of the ester bond conserves a considerable part of the energy from the activation reaction. This stored energy provides the majority of the energy needed for peptide bond formation during translation.

Each of the 20 amino acids are recognized by its specific aminoacyl-tRNA synthetase. The synthetases are usually composed of one to four protein subunits. The enzymes vary considerably in structure although they all perform the same type of reaction by binding ATP, one specific amino acid and its corresponding tRNA.

The specificity of the amino acid activation is as critical for the translational accuracy as the correct matching of the codon with the anticodon. The reason is that the ribosome only sees the anticodon of the tRNA during translation. Thus, the ribosome will not be able to discriminate between tRNAs with the same anticodon but linked to different amino acids.

The error frequency of the amino acid activation reaction is approximately 1 in 10000 despite the small structural differences between some of the amino acids.

Binding to Proteins

Some proteins that bind ATP do so in a characteristic protein fold known as the Rossmann fold, which is a general nucleotide-binding structural domain that can also bind the coenzyme NAD. The most common ATP-binding proteins, known as kinases, share a small number of common folds; the protein kinases, the largest kinase superfamily, all share common structural features specialized for ATP binding and phosphate transfer.

ATP in complexes with proteins, in general, requires the presence of a divalent cation, almost always magnesium, which binds to the ATP phosphate groups. The presence of magnesium greatly decreases the dissociation constant of ATP from its protein binding partner without affecting the ability of the enzyme to catalyze its reaction once the ATP has bound. The presence of magnesium ions can serve as a mechanism for kinase regulation.

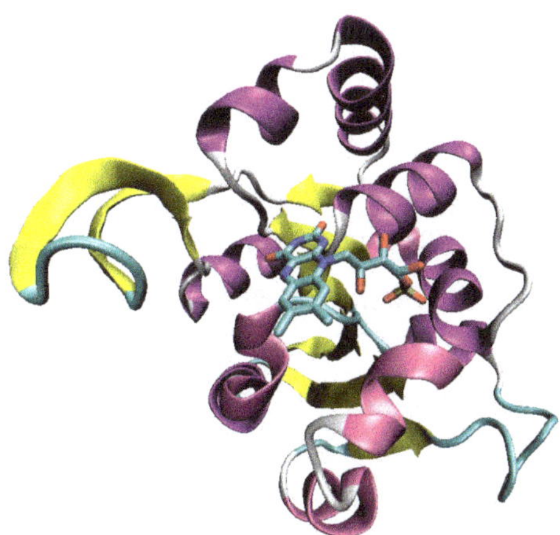

An example of the Rossmann fold, a structural domain of a decarboxylase enzyme from the bacterium Staphylococcus epidermidis (PDB: 1G5Q) with a bound flavin mononucleotide cofactor.

ATP Analogues

Biochemistry laboratories often use *in vitro* studies to explore ATP-dependent molecular processes. Enzyme inhibitors of ATP-dependent enzymes such as kinases are needed to examine the binding sites and transition states involved in ATP-dependent reactions. ATP analogs are also used in X-ray crystallography to determine a protein structure in complex with ATP, often together with other substrates. Most useful ATP analogs cannot be hydrolyzed as ATP would be; instead they trap the enzyme in a structure closely related to the ATP-bound state. Adenosine 5′-(γ-thiotriphosphate) is an extremely common ATP analog in which one of the gamma-phosphate oxygens is replaced by a sulfur atom; this molecule is hydrolyzed at a dramatically slower rate than ATP itself and functions as an inhibitor of ATP-dependent processes. In crystallographic studies, hydrolysis transition states are modeled by the bound vanadate ion. However, caution is warranted in interpreting the results of experiments using ATP analogs, since some enzymes can hydrolyze them at appreciable rates at high concentration.

References

- Murray, Robert K.; Daryl K. Granner; Peter A. Mayes; Victor W. Rodwell (2003). Harper's Illustrated Biochemistry. New York, NY: Lange Medical Books/ MgGraw Hill. p. 96. ISBN 0-07-121766-5.
- Karp, Gerald (2008). Cell and Molecular Biology (5th ed.). Hoboken, NJ: John Wiley & Sons. p. 194. ISBN 978-0-470-04217-5.
- Fenchel T; King GM; Blackburn TH (September 2006). Bacterial Biogeochemistry: The Ecophysiology of Mineral Cycling (2nd ed.). Elsevier. ISBN 978-0-12-103455-9.
- Lengeler JW (January 1999). Drews G; Schlegel HG, eds. Biology of the Prokaryotes. Blackwell Science. ISBN 978-0-632-05357-5.
- Nelson DL; Cox MM (April 2005). Lehninger Principles of Biochemistry (4th ed.). W. H. Freeman. ISBN 978-0-7167-4339-2.

- White D. (September 1999). The Physiology and Biochemistry of Prokaryotes (2nd ed.). Oxford University Press. ISBN 978-0-19-512579-5.
- Heytler PG (1979). "Uncouplers of oxidative phosphorylation". Meth. Enzymol. Methods in Enzymology. 55: 462–42. doi:10.1016/0076-6879(79)55060-5. ISBN 978-0-12-181955-2.
- Campbell, Neil A.; Williamson, Brad; Heyden, Robin J. (2006). Biology: Exploring Life. Boston, MA: Pearson Prentice Hall. ISBN 0-13-250882-6.
- Budavari, Susan, ed. (2001), The Merck Index: An Encyclopedia of Chemicals, Drugs, and Biologicals (13th ed.), Merck, ISBN 0911910131
- Ferguson, S. J.; Nicholls, David; Ferguson, Stuart (2002). Bioenergetics 3 (3rd ed.). San Diego, CA: Academic. ISBN 0-12-518121-3.
- Berg, Jeremy M.; Tymoczko, John L.; Stryer, Lubert (2007). Biochemistry (6th ed.). New York, NY: W. H. Freeman. p. 413. ISBN 0-7167-8724-5.
- Byrnes, Chris (January 5, 2012). "Treating Small Fiber Neuropathy Symptoms With ATP". Working toward Wellness. Retrieved January 6, 2014.

3

Various Fermentation Pathways

There are various organisms that utilize fermentation to produce energy and based on the type of organism and the end products, fermentation can be categorized as lactic acid fermentation, mixed acid fermentation and butanediol fermentation. The content of this section explains these fermentation pathways, the organisms that are responsible for each type and the conditions and metabolites of each type.

Anaerobic Digestion

Anaerobic digestion is a collection of processes by which microorganisms break down biodegradable material in the absence of oxygen. The process is used for industrial or domestic purposes to manage waste or to produce fuels. Much of the fermentation used industrially to produce food and drink products, as well as home fermentation, uses anaerobic digestion.

Anaerobic digestion occurs naturally in some soils and in lake and oceanic basin sediments, where it is usually referred to as "anaerobic activity". This is the source of marsh gas methane as discovered by Volta in 1776.

The digestion process begins with bacterial hydrolysis of the input materials. Insoluble organic polymers, such as carbohydrates, are broken down to soluble derivatives that become available for other bacteria. Acidogenic bacteria then convert the sugars and amino acids into carbon dioxide, hydrogen, ammonia, and organic acids. These bacteria convert these resulting organic acids into acetic acid, along with additional ammonia, hydrogen, and carbon dioxide. Finally, methanogens convert these products to methane and carbon dioxide. The methanogenic archaea populations play an indispensable role in anaerobic wastewater treatments.

Anaerobic digestion is used as part of the process to treat biodegradable waste and sewage sludge. As part of an integrated waste management system, anaerobic digestion reduces the emission of landfill gas into the atmosphere. Anaerobic digesters can also be fed with purpose-grown energy crops, such as maize.

Anaerobic digestion is widely used as a source of renewable energy. The process produces a biogas, consisting of methane, carbon dioxide and traces of other 'contaminant' gases. This biogas can be used directly as fuel, in combined heat and power gas engines or upgraded to natural gas-quality biomethane. The nutrient-rich digestate also produced can be used as fertilizer.

With the re-use of waste as a resource and new technological approaches that have lowered capital costs, anaerobic digestion has in recent years received increased attention among governments in a number of countries, among these the United Kingdom (2011), Germany and Denmark (2011).

Process

Many microorganisms affect anaerobic digestion, including acetic acid-forming bacteria (acetogens) and methane-forming archaea (methanogens). These organisms promote a number of chemical processes in converting the biomass to biogas.

Gaseous oxygen is excluded from the reactions by physical containment. Anaerobes utilize electron acceptors from sources other than oxygen gas. These acceptors can be the organic material itself or may be supplied by inorganic oxides from within the input material. When the oxygen source in an anaerobic system is derived from the organic material itself, the 'intermediate' end products are primarily alcohols, aldehydes, and organic acids, plus carbon dioxide. In the presence of specialised methanogens, the intermediates are converted to the 'final' end products of methane, carbon dioxide, and trace levels of hydrogen sulfide. In an anaerobic system, the majority of the chemical energy contained within the starting material is released by methanogenic bacteria as methane.

Populations of anaerobic microorganisms typically take a significant period of time to establish themselves to be fully effective. Therefore, common practice is to introduce anaerobic microorganisms from materials with existing populations, a process known as "seeding" the digesters, typically accomplished with the addition of sewage sludge or cattle slurry.

Process Stages

The four key stages of anaerobic digestion involve hydrolysis, acidogenesis, acetogenesis and methanogenesis. The overall process can be described by the chemical reaction, where organic material such as glucose is biochemically digested into carbon dioxide (CO_2) and methane (CH_4) by the anaerobic microorganisms.

$$C_6H_{12}O_6 \rightarrow 3CO_2 + 3CH_4$$

- Hydrolysis

In most cases, biomass is made up of large organic polymers. For the bacteria in anaerobic digesters to access the energy potential of the material, these chains must first be broken down into their smaller constituent parts. These constituent parts, or monomers, such as sugars, are readily available to other bacteria. The process of breaking these chains and dissolving the smaller molecules into solution is called hydrolysis. Therefore, hydrolysis of these high-molecular-weight polymeric components is the necessary first step in anaerobic digestion. Through hydrolysis the complex organic molecules are broken down into simple sugars, amino acids, and fatty acids.

Acetate and hydrogen produced in the first stages can be used directly by methanogens. Other molecules, such as volatile fatty acids (VFAs) with a chain length greater than that of acetate must first be catabolised into compounds that can be directly used by methanogens.

- Acidogenesis

The biological process of acidogenesis results in further breakdown of the remaining components by acidogenic (fermentative) bacteria. Here, VFAs are created, along with ammonia, carbon dioxide, and hydrogen sulfide, as well as other byproducts. The process of acidogenesis is similar to the way milk sours.

- Acetogenesis

The third stage of anaerobic digestion is acetogenesis. Here, simple molecules created through the acidogenesis phase are further digested by acetogens to produce largely acetic acid, as well as carbon dioxide and hydrogen.

- Methanogenesis

The terminal stage of anaerobic digestion is the biological process of methanogenesis. Here, methanogens use the intermediate products of the preceding stages and convert them into methane, carbon dioxide, and water. These components make up the majority of the biogas emitted from the system. Methanogenesis is sensitive to both high and low pHs and occurs between pH 6.5 and pH 8. The remaining, indigestible material the microbes cannot use and any dead bacterial remains constitute the digestate.

Configuration

Anaerobic digesters can be designed and engineered to operate using a number of different configurations and can be categorized into batch vs. continuous process mode, mesophilic vs. thermophilic temperature conditions, high vs. low portion of solids, and single stage vs. multistage processes. More initial build money and a larger volume of the batch digester is needed to handle the same amount of waste as a continuous process digester. Higher heat energy is demanded in a thermophilic system compared to a mesophilic system and has a larger gas output capacity and higher methane gas content. For solids content, low will handle up to 15% solid content. Above this level is considered high solids content and can also be known as dry digestion. In a single stage process, one reactor houses the four anaerobic digestion steps. A multistage process utilizes two or more reactors for digestion to separate the methanogenesis and hydrolysis phases.

Batch or Continuous

Anaerobic digestion can be performed as a batch process or a continuous process. In a batch system, biomass is added to the reactor at the start of the process. The reactor is then sealed for the duration of the process. In its simplest form batch processing needs inoculation with already processed material to start the anaerobic digestion. In a typical scenario, biogas production will be formed with a normal distribution pattern over time. Operators can use this fact to determine when they believe the process of digestion of the organic matter has completed. There can be severe odour issues if a batch reactor is opened and emptied before the process is well completed. A more advanced type of batch approach has limited the odour issues by integrating anaerobic digestion with in-vessel composting. In this approach inoculation takes place through the use of recirculated degasified percolate. After anaerobic digestion has completed, the biomass is kept in the reactor which is then used for in-vessel composting before it is opened As the batch digestion is simple and requires less equipment and lower levels of design work, it is typically a cheaper form of digestion. Using more than one batch reactor at a plant can ensure constant production of biogas.

In continuous digestion processes, organic matter is constantly added (continuous complete mixed) or added in stages to the reactor (continuous plug flow; first in – first out). Here, the end products are constantly or periodically removed, resulting in constant production of biogas. A single or multiple digesters in sequence may be used. Examples of this form of anaerobic digestion include continuous stirred-tank reactors, upflow anaerobic sludge blankets, expanded granular sludge beds and internal circulation reactors.

Temperature

The two conventional operational temperature levels for anaerobic digesters determine the species of methanogens in the digesters:

- *Mesophilic* digestion takes place optimally around 30 to 38 °C, or at ambient temperatures between 20 and 45 °C, where mesophiles are the primary microorganism present.
- *Thermophilic* digestion takes place optimally around 49 to 57 °C, or at elevated temperatures up to 70 °C, where thermophiles are the primary microorganisms present.

A limit case has been reached in Bolivia, with anaerobic digestion in temperature working conditions of less than 10 °C. The anaerobic process is very slow, taking more than three times the normal mesophilic time process. In experimental work at University of Alaska Fairbanks, a 1,000 litre digester using psychrophiles harvested from "mud from a frozen lake in Alaska" has produced 200–300 litres of methane per day, about 20 to 30% of the output from digesters in warmer climates. Mesophilic species outnumber thermophiles, and they are also more tolerant to changes in environmental conditions than thermophiles. Mesophilic systems are, therefore, considered to be more stable than thermophilic digestion systems. In contrast, while thermophilic digestion systems are considered less stable, their energy input is higher, with more biogas being removed from the organic matter in an equal amount of time. The increased temperatures facilitate faster reaction rates, and thus faster gas yields. Operation at higher temperatures facilitates greater pathogen reduction of the digestate. In countries where legislation, such as the Animal By-Products Regulations in the European Union, requires digestate to meet certain levels of pathogen reduction there may be a benefit to using thermophilic temperatures instead of mesophilic.

Additional pre-treatment can be used to reduce the necessary retention time to produce biogas. For example, certain processes shred the substrates to increase the surface area or use a thermal pretreatment stage (such as pasteurisation) to significantly enhance the biogas output. The pasteurisation process can also be used to reduce the pathogenic concentration in the digesate leaving the anaerobic digester. Pasteurisation may be achieved by heat treatment combined with maceration of the solids.

Solids Content

In a typical scenario, three different operational parameters are associated with the solids content of the feedstock to the digesters:

- High solids (dry—stackable substrate)
- High solids (wet—pumpable substrate)
- Low solids (wet—pumpable substrate)

High solids (dry) digesters are designed to process materials with a solids content between 25 and 40%. Unlike wet digesters that process pumpable slurries, high solids (dry – stackable substrate) digesters are designed to process solid substrates without the addition of water. The primary styles of dry digesters are continuous vertical plug flow and batch tunnel horizontal digesters. Continuous vertical plug flow digesters are upright, cylindrical tanks where feedstock is continuously fed into the top of the digester, and flows downward by gravity during digestion. In batch tunnel digesters, the feedstock is deposited in tunnel-like chambers with a gas-tight door. Neither approach has mixing inside the digester. The amount of pretreatment, such as contaminant removal, depends both upon the nature of the waste streams being processed and the desired quality of the digestate. Size reduction (grinding) is beneficial in continuous vertical systems, as it accelerates digestion, while batch systems avoid grinding and instead require structure (e.g. yard waste) to reduce compaction of the stacked pile. Continuous vertical dry digesters have a smaller footprint due to the shorter effective retention time and vertical design. Wet digesters can be designed to operate in either a high-solids content, with a total suspended solids (TSS) concentration greater than ~20%, or a low-solids concentration less than ~15%.

High solids (wet) digesters process a thick slurry that requires more energy input to move and process the feedstock. The thickness of the material may also lead to associated problems with abrasion. High solids digesters will typically have a lower land requirement due to the lower volumes associated with the moisture. High solids digesters also require correction of conventional performance calculations (e.g. gas production, retention time, kinetics, etc.) originally based on very dilute sewage digestion concepts, since larger fractions of the feedstock mass are potentially convertible to biogas.

Low solids (wet) digesters can transport material through the system using standard pumps that require significantly lower energy input. Low solids digesters require a larger amount of land than high solids due to the increased volumes associated with the increased liquid-to-feedstock ratio of the digesters. There are benefits associated with operation in a liquid environment, as it enables more thorough circulation of materials and contact between the bacteria and their food. This enables the bacteria to more readily access the substances on which they are feeding, and increases the rate of gas production.

Complexity

Digestion systems can be configured with different levels of complexity. In a single-stage digestion system (one-stage), all of the biological reactions occur within a single, sealed reactor or holding tank. Using a single stage reduces construction costs, but results in less control of the reactions occurring within the system. Acidogenic bacteria, through the production of acids, reduce the pH of the tank. Methanogenic bacteria, as outlined earlier, operate in a strictly defined pH range. Therefore, the biological reactions of the different species in a single-stage reactor can be in direct competition with each other. Another one-stage reaction system is an anaerobic lagoon. These lagoons are pond-like, earthen basins used for the treatment and long-term storage of manures. Here the anaerobic reactions are contained within the natural anaerobic sludge contained in the pool.

In a two-stage digestion system (multistage), different digestion vessels are optimised to bring maximum control over the bacterial communities living within the digesters. Acidogenic bacteria produce organic acids and more quickly grow and reproduce than methanogenic bacteria. Methanogenic bacteria require stable pH and temperature to optimise their performance.

Under typical circumstances, hydrolysis, acetogenesis, and acidogenesis occur within the first reaction vessel. The organic material is then heated to the required operational temperature (either mesophilic or thermophilic) prior to being pumped into a methanogenic reactor. The initial hydrolysis or acidogenesis tanks prior to the methanogenic reactor can provide a buffer to the rate at which feedstock is added. Some European countries require a degree of elevated heat treatment to kill harmful bacteria in the input waste. In this instance, there may be a pasteurisation or sterilisation stage prior to digestion or between the two digestion tanks. Notably, it is not possible to completely isolate the different reaction phases, and often some biogas is produced in the hydrolysis or acidogenesis tanks.

Residence Time

The residence time in a digester varies with the amount and type of feed material, and with the configuration of the digestion system. In a typical two-stage mesophilic digestion, residence time varies between 15 and 40 days, while for a single-stage thermophilic digestion, residence times is normally faster and takes around 14 days. The plug-flow nature of some of these systems will mean the full degradation of the material may not have been realised in this timescale. In this event, digestate exiting the system will be darker in colour and will typically have more odour.

In the case of an upflow anaerobic sludge blanket digestion (UASB), hydraulic residence times can be as short as 1 hour to 1 day, and solid retention times can be up to 90 days. In this manner, a UASB system is able to separate solids and hydraulic retention times with the use of a sludge blanket. Continuous digesters have mechanical or hydraulic devices, depending on the level of solids in the material, to mix the contents, enabling the bacteria and the food to be in contact. They also allow excess material to be continuously extracted to maintain a reasonably constant volume within the digestion tanks.

Inhibition

The anaerobic digestion process can be inhibited by several compounds, affecting one or more of the bacterial groups responsible for the different organic matter degradation steps. The degree of the inhibition depends, among other factors, on the concentration of the inhibitor in the digester. Potential inhibitors are ammonia, sulfide, light metal ions (Na, K, Mg, Ca, Al), heavy metals, some organics (chlorophenols, halogenated aliphatics, N-substituted aromatics, long chain fatty acids), etc.

Left: Farm-based maize silage digester located near Neumünster in Germany, 2007 - the green, inflatable biogas holder is shown on top of the digester. *Right:* Two-stage, low solids, UASB digestion component of a mechanical biological treatment system near Tel Aviv; the process water is seen in balance tank and sequencing batch reactor, 2005.

Feedstocks

The most important initial issue when considering the application of anaerobic digestion systems is the feedstock to the process. Almost any organic material can be processed with anaerobic digestion; however, if biogas production is the aim, the level of putrescibility is the key factor in its successful application. The more putrescible (digestible) the material, the higher the gas yields possible from the system.

Anaerobic lagoon and generators at the Cal Poly Dairy, United States

Feedstocks can include biodegradable waste materials, such as waste paper, grass clippings, leftover food, sewage, and animal waste. Woody wastes are the exception, because they are largely unaffected by digestion, as most anaerobes are unable to degrade lignin. Xylophalgeous anaerobes (lignin consumers) or using high temperature pretreatment, such as pyrolysis, can be used to break down the lignin. Anaerobic digesters can also be fed with specially grown energy crops, such as silage, for dedicated biogas production. In Germany and continental Europe, these facilities are referred to as "biogas" plants. A codigestion or cofermentation plant is typically an agricultural anaerobic digester that accepts two or more input materials for simultaneous digestion.

The length of time required for anaerobic digestion depends on the chemical complexity of the material. Material rich in easily digestible sugars breaks down quickly where as intact lignocellulosic material rich in cellulose and hemicellulose polymers can take much longer to break down. Anaerobic microorganisms are generally unable to break down lignin, the recalcitrant aromatic component of biomass.

Anaerobic digesters were originally designed for operation using sewage sludge and manures. Sewage and manure are not, however, the material with the most potential for anaerobic digestion, as the biodegradable material has already had much of the energy content taken out by the animals that produced it. Therefore, many digesters operate with codigestion of two or more types of feedstock. For example, in a farm-based digester that uses dairy manure as the primary feedstock, the gas production may be significantly increased by adding a second feedstock, e.g., grass and corn (typical on-farm feedstock), or various organic byproducts, such as slaughterhouse waste, fats, oils and grease from restaurants, organic household waste, etc. (typical off-site feedstock).

Digesters processing dedicated energy crops can achieve high levels of degradation and biogas production. Slurry-only systems are generally cheaper, but generate far less energy than those using crops, such as maize and grass silage; by using a modest amount of crop material (30%), an anaerobic digestion plant can increase energy output tenfold for only three times the capital cost, relative to a slurry-only system.

Moisture Content

A second consideration related to the feedstock is moisture content. Dryer, stackable substrates, such as food and yard waste, are suitable for digestion in tunnel-like chambers. Tunnel-style systems typically have near-zero wastewater discharge, as well, so this style of system has advantages where the discharge of digester liquids are a liability. The wetter the material, the more suitable it will be to handling with standard pumps instead of energy-intensive concrete pumps and physical means of movement. Also, the wetter the material, the more volume and area it takes up relative to the levels of gas produced. The moisture content of the target feedstock will also affect what type of system is applied to its treatment. To use a high-solids anaerobic digester for dilute feedstocks, bulking agents, such as compost, should be applied to increase the solids content of the input material. Another key consideration is the carbon:nitrogen ratio of the input material. This ratio is the balance of food a microbe requires to grow; the optimal C:N ratio is 20–30:1. Excess N can lead to ammonia inhibition of digestion.

Contamination

The level of contamination of the feedstock material is a key consideration. If the feedstock to the digesters has significant levels of physical contaminants, such as plastic, glass, or metals, then processing to remove the contaminants will be required for the material to be used. If it is not removed, then the digesters can be blocked and will not function efficiently. It is with this understanding that mechanical biological treatment plants are designed. The higher the level of pretreatment a feedstock requires, the more processing machinery will be required, and, hence, the project will have higher capital costs.

After sorting or screening to remove any physical contaminants from the feedstock, the material is often shredded, minced, and mechanically or hydraulically pulped to increase the surface area available to microbes in the digesters and, hence, increase the speed of digestion. The maceration of solids can be achieved by using a chopper pump to transfer the feedstock material into the airtight digester, where anaerobic treatment takes place.

Substrate Composition

Substrate composition is a major factor in determining the methane yield and methane production rates from the digestion of biomass. Techniques to determine the compositional characteristics of the feedstock are available, while parameters such as solids, elemental, and organic analyses are important for digester design and operation.

Applications

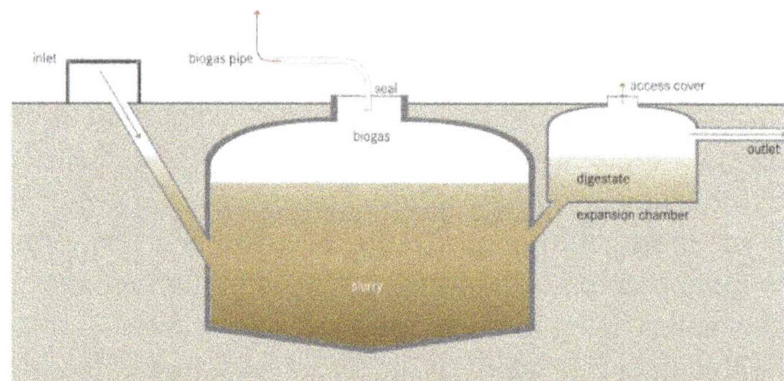

Schematic of an anaerobic digester as part of a sanitation system. It produces a digested slurry (digestate) that can be used as a fertilizer, and biogas that can be used for energy.

Using anaerobic digestion technologies can help to reduce the emission of greenhouse gases in a number of key ways:

- Replacement of fossil fuels
- Reducing or eliminating the energy footprint of waste treatment plants
- Reducing methane emission from landfills
- Displacing industrially produced chemical fertilizers
- Reducing vehicle movements
- Reducing electrical grid transportation losses
- Reducing usage of LP Gas for cooking

Waste and Wastewater Treatment

Anaerobic digestion is particularly suited to organic material, and is commonly used for industrial effluent, wastewater and sewage sludge treatment. Anaerobic digestion, a simple process, can greatly reduce the amount of organic matter which might otherwise be destined to be dumped at sea, dumped in landfills, or burnt in incinerators.

Pressure from environmentally related legislation on solid waste disposal methods in developed countries has increased the application of anaerobic digestion as a process for reducing waste volumes and generating useful byproducts. It may either be used to process the source-separated fraction of municipal waste or alternatively combined with mechanical sorting systems, to process residual mixed municipal waste. These facilities are called mechanical biological treatment plants.

If the putrescible waste processed in anaerobic digesters were disposed of in a landfill, it would break down naturally and often anaerobically. In this case, the gas will eventually escape into the atmosphere. As methane is about 20 times more potent as a greenhouse gas than carbon dioxide, this has significant negative environmental effects.

In countries that collect household waste, the use of local anaerobic digestion facilities can help to reduce the amount of waste that requires transportation to centralized landfill sites or incineration facilities. This reduced burden on transportation reduces carbon emissions from the collection vehicles. If localized anaerobic digestion facilities are embedded within an electrical distribution network, they can help reduce the electrical losses associated with transporting electricity over a national grid.

Power Generation

In developing countries, simple home and farm-based anaerobic digestion systems offer the potential for low-cost energy for cooking and lighting. From 1975, China and India have both had large, government-backed schemes for adaptation of small biogas plants for use in the household for cooking and lighting. At present, projects for anaerobic digestion in the developing world can gain financial support through the United Nations Clean Development Mechanism if they are able to show they provide reduced carbon emissions.

Methane and power produced in anaerobic digestion facilities can be used to replace energy derived from fossil fuels, and hence reduce emissions of greenhouse gases, because the carbon in biodegradable material is part of a carbon cycle. The carbon released into the atmosphere from the combustion of biogas has been removed by plants for them to grow in the recent past, usually within the last decade, but more typically within the last growing season. If the plants are regrown, taking the carbon out of the atmosphere once more, the system will be carbon neutral. In contrast, carbon in fossil fuels has been sequestered in the earth for many millions of years, the combustion of which increases the overall levels of carbon dioxide in the atmosphere.

Biogas from sewage works is sometimes used to run a gas engine to produce electrical power, some or all of which can be used to run the sewage works. Some waste heat from the engine is then used to heat the digester. The waste heat is, in general, enough to heat the digester to the required temperatures. The power potential from sewage works is limited – in the UK, there are about 80 MW total of such generation, with the potential to increase to 150 MW, which is insignificant compared to the average power demand in the UK of about 35,000 MW. The scope for biogas generation from nonsewage waste biological matter – energy crops, food waste, abattoir waste, etc. - is much higher, estimated to be capable of about 3,000 MW. Farm biogas plants using animal waste and energy crops are expected to contribute to reducing CO_2 emissions and strengthen the grid, while providing UK farmers with additional revenues.

Some countries offer incentives in the form of, for example, feed-in tariffs for feeding electricity onto the power grid to subsidize green energy production.

In Oakland, California at the East Bay Municipal Utility District's main wastewater treatment plant (EBMUD), food waste is currently codigested with primary and secondary municipal wastewater solids and other high-strength wastes. Compared to municipal wastewater solids digestion alone, food waste codigestion has many benefits. Anaerobic digestion of food waste pulp from the EBMUD food waste process provides a higher normalized energy benefit, compared to municipal wastewater solids: 730 to 1,300 kWh per dry ton of food waste applied compared to 560 to 940 kWh per dry ton of municipal wastewater solids applied.

Grid Injection

Biogas grid-injection is the injection of biogas into the natural gas grid. The raw biogas has to be previously upgraded to biomethane. This upgrading implies the removal of contaminants such as hydrogen sulphide or siloxanes, as well as the carbon dioxide. Several technologies are available for this purpose, being the most widely implemented the pressure swing adsorption (PSA), water or amine scrubbing (absorption processes) and, in the last years, membrane separation. As an alternative, the electricity and the heat can be used for on-site generation, resulting in a reduction of losses in the transportation of energy. Typical energy losses in natural gas transmission systems range from 1–2%, whereas the current energy losses on a large electrical system range from 5–8%.

In October 2010, Didcot Sewage Works became the first in the UK to produce biomethane gas supplied to the national grid, for use in up to 200 homes in Oxfordshire. By 2017, UK electricity firm Ecotricity plan to have digester fed by locally sourced grass fueling 6000 homes

Vehicle Fuel

After upgrading with the above-mentioned technologies, the biogas (transformed into biomethane) can be used as vehicle fuel in adapted vehicles. This use is very extensive in Sweden, where over 38,600 gas vehicles exist, and 60% of the vehicle gas is biomethane generated in anaerobic digestion plants.

Fertiliser and Soil Conditioner

The solid, fibrous component of the digested material can be used as a soil conditioner to increase the organic content of soils. Digester liquor can be used as a fertiliser to supply vital nutrients to soils instead of chemical fertilisers that require large amounts of energy to produce and transport. The use of manufactured fertilisers is, therefore, more carbon-intensive than the use of anaerobic digester liquor fertiliser. In countries such as Spain, where many soils are organically depleted, the markets for the digested solids can be equally as important as the biogas.

Cooking Gas

By using a bio-digester, which produces the bacteria required for decomposing, cooking gas is generated. The organic garbage like fallen leaves, kitchen waste, food waste etc. are fed into a crusher unit, where the mixture is conflated with a small amount of water. The mixture is then fed into the bio-digester, where the bacteria decomposes it to produce cooking gas. This gas is piped to kitchen stove. A 2 cubic meter bio-digester can produce 2 cubic meter of cooking gas. This is equivalent to 1 kg of LPG. The notable advantage of using a bio-digester is the sludge which is a rich organic manure.

Products

The three principal products of anaerobic digestion are biogas, digestate, and water.

Biogas

Typical composition of biogas		
Compound	**Formula**	**%**
Methane	CH4	50–75
Carbon dioxide	CO2	25–50
Nitrogen	N2	0–10
Hydrogen	H2	0–1
Hydrogen sulphide	H2S	0–3
Oxygen	O2	0–0
Source: *www.kolumbus.fi, 2007*		

Biogas is the ultimate waste product of the bacteria feeding off the input biodegradable feedstock (the methanogenesis stage of anaerobic digestion is performed by archaea (a micro-organism on a distinctly different branch of the phylogenetic tree of life to bacteria), and is mostly methane and carbon dioxide, with a small amount hydrogen and trace hydrogen sulfide. (As-produced, biogas also contains water vapor, with the fractional water vapor volume a function of biogas temperature). Most of the biogas is produced during the middle of the digestion, after the bacterial population has grown, and tapers off as the putrescible material is exhausted. The gas is normally stored on top of the digester in an inflatable gas bubble or extracted and stored next to the facility in a gas holder.

The methane in biogas can be burned to produce both heat and electricity, usually with a reciprocating engine or microturbine often in a cogeneration arrangement where the electricity and waste heat generated are used to warm the digesters or to heat buildings. Excess electricity can be sold to suppliers or put into the local grid. Electricity produced by anaerobic digesters is considered to be renewable energy and may attract subsidies. Biogas does not contribute to increasing atmospheric carbon dioxide concentrations because the gas is not released directly into the atmosphere and the carbon dioxide comes from an organic source with a short carbon cycle.

Biogas may require treatment or 'scrubbing' to refine it for use as a fuel. Hydrogen sulfide, a toxic product formed from sulfates in the feedstock, is released as a trace component of the biogas. National environmental enforcement agencies, such as the U.S. Environmental Protection Agency or the English and Welsh Environment Agency, put strict limits on the levels of gases containing hydrogen sulfide, and, if the levels of hydrogen sulfide in the gas are high, gas scrubbing and cleaning equipment (such as amine gas treating) will be needed to process the biogas to within regionally accepted levels. Alternatively, the addition of ferrous chloride $FeCl_2$ to the digestion tanks inhibits hydrogen sulfide production.

Volatile siloxanes can also contaminate the biogas; such compounds are frequently found in household waste and wastewater. In digestion facilities accepting these materials as a component of the feedstock, low-molecular-weight siloxanes volatilise into biogas. When this gas is combusted in a gas engine, turbine, or boiler, siloxanes are converted into silicon dioxide (SiO_2), which deposits internally in the machine, increasing wear and tear. Practical and cost-effective technologies to remove siloxanes and other biogas contaminants are available at the present time. In certain applications, *in situ* treatment can be used to increase the methane purity by reducing the offgas carbon dioxide content, purging the majority of it in a secondary reactor.

In countries such as Switzerland, Germany, and Sweden, the methane in the biogas may be compressed for it to be used as a vehicle transportation fuel or input directly into the gas mains. In countries where the driver for the use of anaerobic digestion are renewable electricity subsidies, this route of treatment is less likely, as energy is required in this processing stage and reduces the overall levels available to sell.

Biogas holder with lightning protection rods and backup gas flare

Biogas carrying pipes

Digestate

Digestate is the solid remnants of the original input material to the digesters that the microbes cannot use. It also consists of the mineralised remains of the dead bacteria from within the digesters. Digestate can come in three forms: fibrous, liquor, or a sludge-based combination of the two fractions. In two-stage systems, different forms of digestate come from different digestion tanks. In single-stage digestion systems, the two fractions will be combined and, if desired, separated by further processing.

Acidogenic anaerobic digestate

The second byproduct (acidogenic digestate) is a stable, organic material consisting largely of lignin and cellulose, but also of a variety of mineral components in a matrix of dead bacterial cells; some plastic may be present. The material resembles domestic compost and can be used as such or to make low-grade building products, such as fibreboard. The solid digestate can also be used as feedstock for ethanol production.

The third byproduct is a liquid (methanogenic digestate) rich in nutrients, which can be used as a fertiliser, depending on the quality of the material being digested. Levels of potentially toxic elements (PTEs) should be chemically assessed. This will depend upon the quality of the original feedstock. In the case of most clean and source-separated biodegradable waste streams, the levels of PTEs will be low. In the case of wastes originating from industry, the levels of PTEs may be higher and will need to be taken into consideration when determining a suitable end use for the material.

Digestate typically contains elements, such as lignin, that cannot be broken down by the anaerobic microorganisms. Also, the digestate may contain ammonia that is phytotoxic, and may hamper the growth of plants if it is used as a soil-improving material. For these two reasons, a maturation or composting stage may be employed after digestion. Lignin and other materials are available for degradation by aerobic microorganisms, such as fungi, helping reduce the overall volume of the material for transport. During this maturation, the ammonia will be oxidized into nitrates, improving the fertility of the material and making it more suitable as a soil improver. Large composting stages are typically used by dry anaerobic digestion technologies.

Wastewater

The final output from anaerobic digestion systems is water, which originates both from the moisture content of the original waste that was treated and water produced during the microbial reactions in the digestion systems. This water may be released from the dewatering of the digestate or may be implicitly separate from the digestate.

The wastewater exiting the anaerobic digestion facility will typically have elevated levels of biochemical oxygen demand (BOD) and chemical oxygen demand (COD). These measures of the reactivity of the effluent indicate an ability to pollute. Some of this material is termed 'hard COD', meaning it cannot be accessed by the anaerobic bacteria for conversion into biogas. If this effluent were put directly into watercourses, it would negatively affect them by causing eutrophication. As such, further treatment of the wastewater is often required. This treatment will typically be an oxidation stage wherein air is passed through the water in a sequencing batch reactors or reverse osmosis unit.

History

Reported scientific interest in the manufacturing of gas produced by the natural decomposition of organic matter dates from the 17th century, when Robert Boyle (1627-1691) and Stephen Hales (1677-1761) noted that disturbing the sediment of streams and lakes released flammable gas. In 1808 Sir Humphry Davy proved the presence of methane in the gases produced by cattle manure. In 1859 a leper colony in Bombay in India built the first anaerobic digester. In 1895, the technology was developed in Exeter, England, where a septic tank was used to generate gas for the sewer gas destructor lamp, a type of gas lighting. Also in England, in 1904, the first dual-purpose tank for both sedimentation and sludge treatment was installed in Hampton, London. In 1907, in Germany, a patent was issued for the Imhoff tank, an early form of digester.

Gas street lamp

Research on anaerobic digestion began in earnest in the 1930s.

Lactic Acid Fermentation

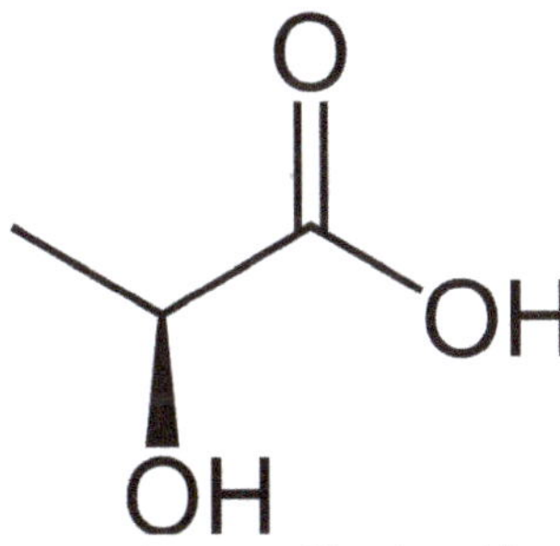

One isomer of lactic acid

Lactic acid fermentation is a metabolic process by which glucose and other six-carbon sugars (also, disaccharides of six-carbon sugars, e.g. sucrose or lactose) are converted into cellular energy and the metabolite lactate. It is an anaerobic fermentation reaction that occurs in some bacteria and animal cells, such as muscle cells.

If oxygen is present in the cell, many organisms will bypass fermentation and undergo cellular respiration; however, facultative anaerobic organisms will both ferment and undergo respiration in the presence of oxygen. Sometimes even when oxygen is present and aerobic metabolism is happening in the mitochondria, if pyruvate is building up faster than it can be metabolized, the fermentation will happen anyway.

Lactate dehydrogenase catalyzes the interconversion of pyruvate and lactate with concomitant interconversion of NADH and NAD^+.

In *homolactic fermentation*, one molecule of glucose is ultimately converted to two molecules of lactic acid. *Heterolactic fermentation*, in contrast, yields carbon dioxide and ethanol in addition to lactic acid, in a process called the phosphoketolase pathway.

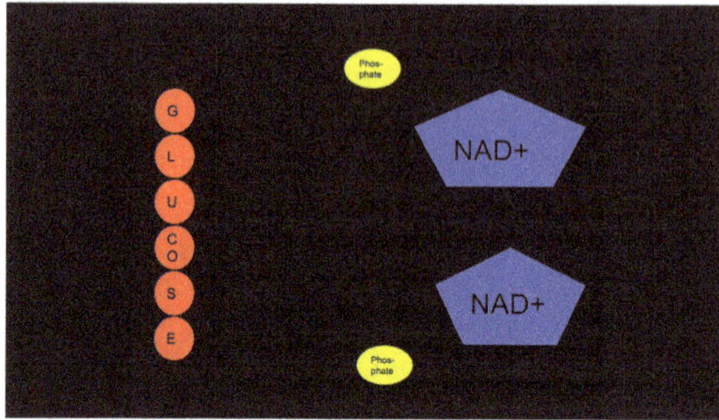

This animation focuses on one molecule of glucose turning into pyruvate then into lactic acid.In the process there is one 6 carbon glucose molecule and 2 NAD+ molecules. 2 phosphates attach to the ends of the glucose molecule, splitting the glucose molecule into 2 pyruvate molecules. The NAD+ molecules add another phosphate onto the open ends of the 2 pyruvate molecules, turning the 2 NAD+ into 2 NADP. Then ADP comes and takes the phosphates, creating 2 ATP molecules.The pyruvate is turned into 2 lactate molecules.The process then repeats, starting with another glucose molecule.

Applications

Lactic acid fermentation is used in many areas of the world to produce foods that cannot be produced through other methods. The most commercially important genus of lactic acid-fermenting bacteria is *Lactobacillus*, though other bacteria and even yeast are sometimes used. Two of the most common applications of lactic acid fermentation are in the production of yogurt and sauerkraut.

Kimchi

Kimchi also uses lactic acid fermentation.

Sauerkraut

Lactic acid fermentation is also used in the production of sauerkraut. The main type of bacteria used in the production of sauerkraut is of the genus *Leuconostoc*.

As in yogurt, when the acidity rises due to lactic acid-fermenting organisms, many other pathogenic microorganisms are killed. The bacteria produce lactic acid, as well as simple alcohols and other hydrocarbons. These may then combine to form esters, contributing to the unique flavor of sauerkraut.

Sour Beer

Lactic acid is a component in the production of sour beers, including Lambics and Berliner Weisses.

Yogurt

The main method of producing yogurt is through the lactic acid fermentation of milk with harmless bacteria. The primary bacteria used are typically *Lactobacillus bulgaricus* and *Streptococcus thermophilus*, and United States law requires all yogurts to contain these two cultures (though others may be added as probiotic cultures). These bacteria produce lactic acid in the milk culture, decreasing its pH and causing it to congeal. The bacteria also produce compounds that give yogurt its distinctive flavor. An additional effect of the lowered pH is the incompatibility of the acidic environment with many other types of harmful bacteria.

For a probiotic yogurt, additional types of bacteria such as *Lactobacillus acidophilus* are also added to the culture.

Physiological

Lactobacillus fermentation and accompanying production of acid provides a protective vaginal microbiome that protects against the proliferation of pathogenic organisms.

Mixed Acid Fermentation

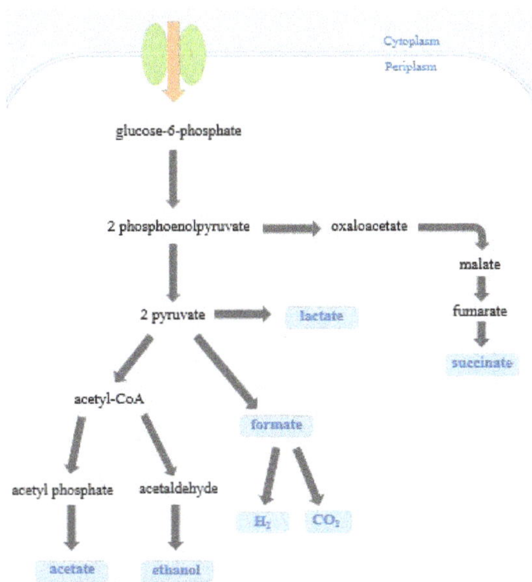

The Mixed Acid Fermentation Pathway in E. coli. End products are highlighted in blue.

Mixed acid fermentation is the biological process by which a six-carbon sugar e.g. glucose is converted into a complex and variable mixture of acids. It is an anaerobic fermentation reaction that is common in bacteria. It is characteristic for members of the Enterobacteriaceae, a large family of Gram-negative bacteria that includes *E. coli*.

The mixture of end products produced by mixed acid fermentation includes lactate, acetate, succinate, formate, ethanol and the gases H_2 and CO_2. The formation of these end products depends on the presence of certain key enzymes in the bacterium. The proportion in which they are formed varies between different bacterial species. The mixed acid fermentation pathway differs from other fermentation pathways, which produce fewer end products in fixed amounts.

The end products of mixed acid fermentation can have many useful applications in biotechnology and industry. For instance, ethanol is widely used as a biofuel. Therefore, multiple bacterial strains have been metabolically engineered in the laboratory to increase the individual yields of certain end products. This research has been carried out primarily in *E. coli* and is ongoing.

Mixed Acid Fermentation in E. Coli

E. coli use fermentation pathways as a final option for energy metabolism, as they produce very little energy in comparison to respiration. Mixed acid fermentation in *E. coli* occurs in two stages. These stages are outlined by the biological database for *E. coli*, EcoCyc.

The first of these two stages is a glycolysis reaction. Under anaerobic conditions, a glycolysis reaction takes place where glucose is converted into pyruvate:

glucose → 2 pyruvate

There is a net production of 2 ATP and 2 NADH molecules per molecule of glucose converted. ATP is generated by substrate-level phosphorylation. NADH is formed from the reduction of NAD.

In the second stage, pyruvate produced by glycolysis is converted to one or more end products via the following reactions. In each case, both of the NADH molecules generated by glycolysis are reoxidized to NAD^+. Each alternative pathway requires a different key enzyme in *E. coli*. After the variable amounts of different end products are formed by these pathways, they are secreted from the cell.

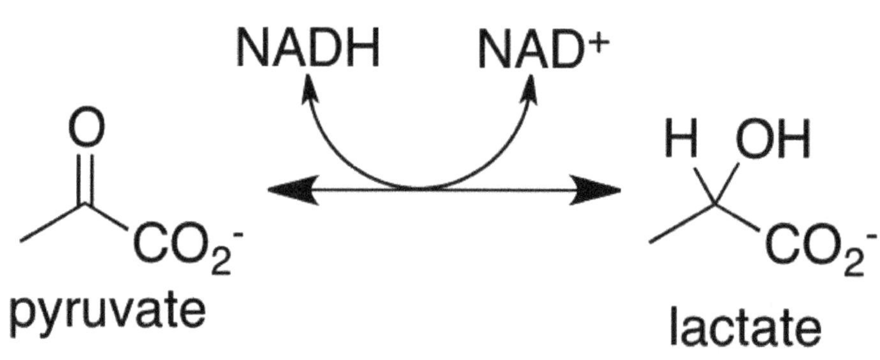

The conversion of pyruvate to lactate is catalysed by the enzyme lactate dehydrogenase.

Lactate Formation

Pyruvate produced by glycolysis is converted to lactate. This reaction is catalysed by the enzyme lactate dehydrogenase (LDHA).

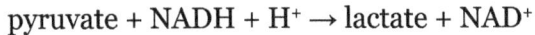

pyruvate + NADH + H^+ → lactate + NAD^+

Acetate Formation

Pyruvate is converted into acetyl-coenzyme A (acetyl-CoA) by the enzyme pyruvate dehydrogenase. This acetyl-CoA is then converted into acetate in *E. coli*, whilst producing ATP by substrate-level phosphorylation. Acetate formation requires two enzymes: phosphate acetyltransferase and acetate kinase.

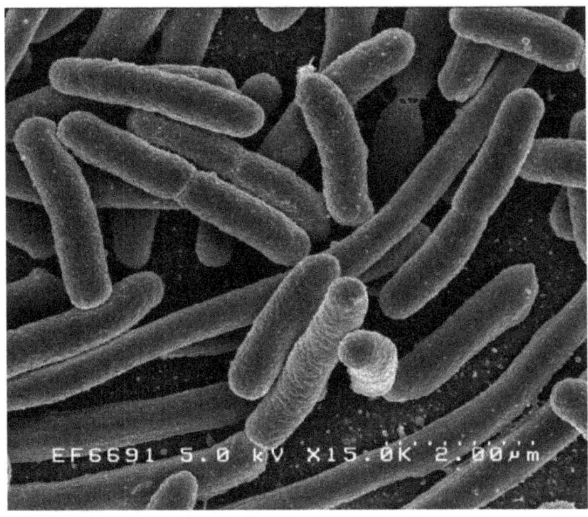

The Mixed Acid Fermentation pathway is characteristic of the Enterobacteriaceae family that includes E. coli

acetyl-CoA + phosphate → acetyl-phosphate + CoA

acetyl-phosphate + ADP → acetate + ATP

Ethanol Formation

Ethanol is formed in *E. coli* by the reduction of acetyl coenzyme A using NADH. This two-step reaction requires the enzyme alcohol dehydrogenase (ADHE).

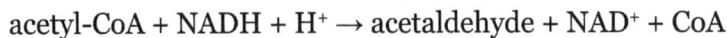

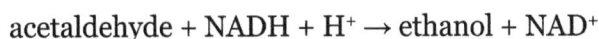

acetyl-CoA + NADH + H^+ → acetaldehyde + NAD^+ + CoA

acetaldehyde + NADH + H^+ → ethanol + NAD^+

Formate Formation

Formate is produced by the cleavage of pyruvate. This reaction is catalysed by the enzyme pyruvate-formate lyase (PFL), which plays an important role in regulating anaerobic fermentation in *E. coli.*

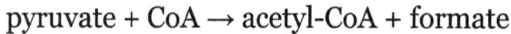

pyruvate + CoA → acetyl-CoA + formate

Succinate Formation

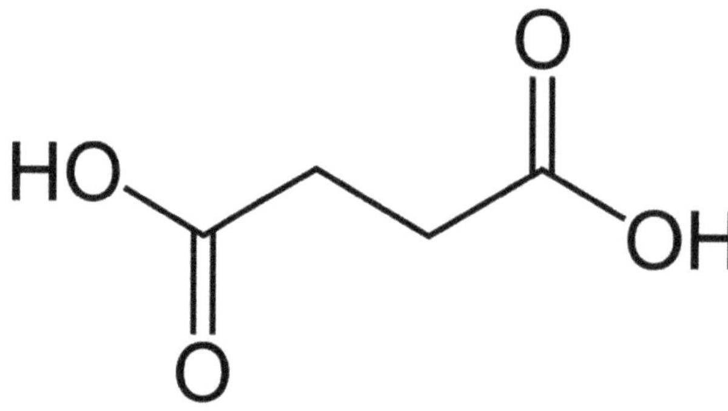

Skeletal structure of succinate

Succinate is formed in *E. coli* in several steps.

Phosphoenolpyruvate (PEP), a glycolysis pathway intermediate, is carboxylated by the enzyme PEP carboxylase to form oxaloacetate. This is followed by the conversion of oxaloacetate to malate by the enzyme malate dehydrogenase. Fumarate hydratase then catalyses the dehydration of malate to produce fumarate.

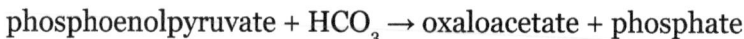

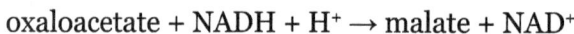

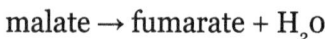

phosphoenolpyruvate + HCO_3 → oxaloacetate + phosphate

oxaloacetate + NADH + H^+ → malate + NAD^+

malate → fumarate + H_2O

The final reaction in the formation of succinate is the reduction of fumarate. It is catalysed by the enzyme fumarate reductase.

fumarate + NADH + H^+ → succinate + NAD^+

This reduction is an anaerobic respiration reaction in *E. coli,* as it uses electrons associated with NADH dehydrogenase and the electron transport chain. ATP is generated by using an electrochemical gradient and ATP synthase. This is the only case in the mixed acid fermentation pathway where ATP is not produced via substrate-level phosphorylation.

Vitamin K_2, also known as menaquinone, is very important for electron transport to fumarate in *E. coli.*

Hydrogen and Carbon Dioxide Formation

Formate can be converted to hydrogen gas and carbon dioxide in *E. coli.* This reaction requires the enzyme formate-hydrogen lyase. It can be used to prevent the conditions inside the cell becoming too acidic.

formate → H_2 and Co_2

Methyl Red Test

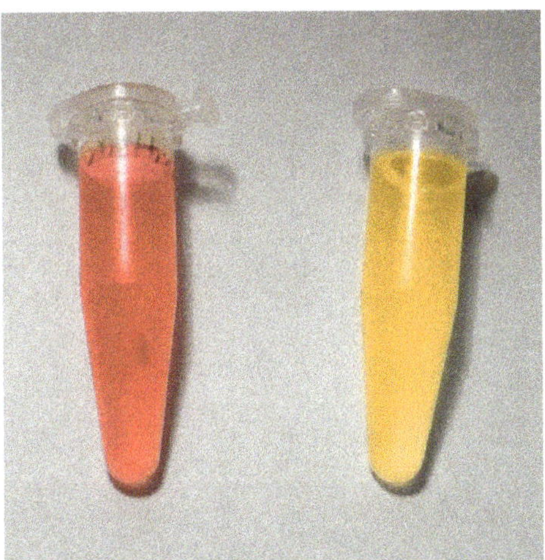

Methyl Red Test: The test tube on the left shows a positive result as acidic end products are formed by mixed acid fermentation in *E. coli*. The test tube on the right shows a negative result as no acidic products are formed by fermentation.

The methyl red (MR) test can detect whether the mixed acid fermentation pathway occurs in microbes when given glucose. A pH indicator is used that turns the test solution red if the pH drops below 4.4. If the fermentation pathway has taken place, the mixture of acids it has produced will make the solution very acidic and cause a red colour change.

The Methyl red test belongs to a group known as the IMViC tests.

Metabolic Engineering

Multiple bacterial strains have been metabolically engineered to increase the individual yields of end products formed by mixed acid fermentation. For instance, strains for the increased production of ethanol, lactate, succinate and acetate have been developed due to the usefulness of these products in biotechnology. The major limiting factor for this engineering is the need to maintain a redox balance in the mixture of acids produced by the fermentation pathway.

For Ethanol Production

Ethanol is the most commonly used biofuel and can be produced on large scale via fermentation. The maximum theoretical yield for the production of ethanol was achieved around 20 years. A plasmid that carried the pyruvate decarboxylase and alcohol dehydrogenase genes from the bacteria *Z.mobilis* was used by scientists. This was inserted into *E. coli* and resulted in an increased yield of ethanol. The genome of this *E. coli* strain, KO11, has more recently been sequenced and mapped.

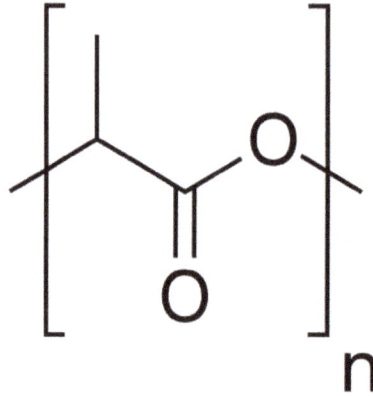

The skeletal formula of Polylactid acid

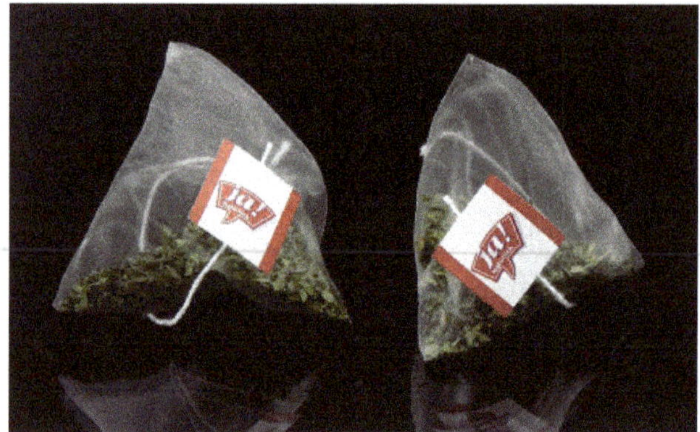

Teabags made from Polylactic acid (PLA)

For Acetate Production

The *E. coli* strain W3110 was genetically engineered to generate 2 moles of acetate for every 1 mole of glucose that undergoes fermentation. This is known as a homoacetate pathway.

For Lactate Production

Lactate can be used to produce a bioplastic called Polylactic acid (PLA). The properties of PLA depend on the ratio of the two optical isomers of lactate (D-lactate and L-lactate). D-lactate is produced by mixed acid fermentation in *E. coli.* Early experiments engineered the E.coli strain RR1 to produce either one of the two optical isomers of lactate.

Later experiments modified the *E. coli strain* KO11, originally developed to enhance ethanol production. Scientists were able to increase the yield of D-lactate from fermentation by performing several deletions.

For Succinate Production

Increasing the yield of succinate from mixed acid fermentation was first done by overexpressing the enzyme PEP carboxylase. This produced a succinate yield that was approximately 3 times greater than normal. Several experiments using a similar approach have followed.

Alternative approaches have altered the redox and ATP balance to optimize the succinate yield.

Butanediol Fermentation

2,3-Butanediol fermentation is anaerobic fermentation of glucose with 2,3-butanediol as one of the end products. The overall stoichiometry of the reaction is

2 pyruvate + NADH --> 2CO_2 + 2,3-butanediol.

Butanediol fermentation is typical for *Klebsiella* and *Enterobacter* and is tested for using the Voges–Proskauer (VP) test.

The metabolic function of 2,3-butanediol is not known.

Comparison with Mixed Acid Fermentation

2,3-butanediol fermentation produces smaller amounts of acid than mixed acid fermentation, and butanediol, ethanol, CO_2 and H_2 are the end products. While equal amounts of CO_2 and H_2 are created during mixed acid fermentation, butanediol fermentation produces more than twice the amount of CO_2 because the gases are not produced only by formate hydrogen lyase like they are in the mixed acid fermentation

Ethanol Fermentation

Ethanol fermentation, also called alcoholic fermentation, is a biological process which converts sugars such as glucose, fructose, and sucrose into cellular energy, producing ethanol and carbon dioxide as a side-effect. Because yeasts perform this conversion in the absence of oxygen, alcoholic fermentation is considered an anaerobic process.

Ethanol fermentation has many uses, including the production of alcoholic beverages, the production of ethanol fuel, and bread baking.

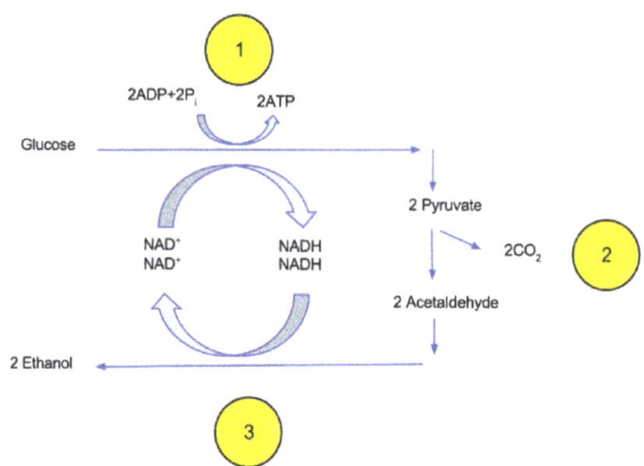

In ethanol fermentation, (1) one glucose molecule breaks down into two pyruvates. The energy from this exothermic reaction is used to bind inorganic phosphates to ADP and convert NAD+ to NADH. (2) The two pyruvates are then broken down into two acetaldehydes and give off two CO2 as a waste product. (3) The two acetaldehydes are then converted to two ethanol by using the H- ions from NADH, converting NADH back into NAD+.

Biochemical Process of Fermentation of Sucrose

A laboratory vessel being used for the fermentation of straw.

The chemical equations below summarize the fermentation of sucrose ($C_{12}H_{22}O_{11}$) into ethanol (C_2H_5OH). Alcoholic fermentation converts one mole of glucose into two moles of ethanol and two moles of carbon dioxide, producing two moles of ATP in the process.

The overall chemical formula for alcoholic fermentation is:

$$C_6H_{12}O_6 \rightarrow 2\ C_2H_5OH + 2\ CO_2$$

Sucrose is a dimer of glucose and fructose molecules. In the first step of alcoholic fermentation, the enzyme invertase cleaves the glycosidic linkage between the glucose and fructose molecules.

$$C_{12}H_{22}O_{11} + H_2O + \text{invertase} \rightarrow 2\ C_6H_{12}O_6$$

Next, each glucose molecule is broken down into two pyruvate molecules in a process known as glycolysis. Glycolysis is summarized by the equation:

$$C_6H_{12}O_6 + 2\ ADP + 2\ P_i + 2\ NAD^+ \rightarrow 2\ CH_3COCOO^- + 2\ ATP + 2\ NADH + 2\ H_2O + 2\ H^+$$

The chemical formula of pyruvate is CH_3COCOO^-. P_i stands for the inorganic phosphate.

Finally, pyruvate is converted to ethanol and CO_2 in two steps, regenerating oxidized NAD+ needed for glycolysis:

1. $CH_3COCOO^- + H^+ \rightarrow CH_3CHO + CO_2$

catalyzed by pyruvate decarboxylase

2. $CH_3CHO + NADH{+}H^+ \rightarrow C_2H_5OH + NAD^+$

This reaction is catalyzed by alcohol dehydrogenase (ADH1 in baker's yeast).

As shown by the reaction equation, glycolysis causes the reduction of two molecules of NAD^+ to NADH. Two ADP molecules are also converted to two ATP and two water molecules via substrate-level phosphorylation.

Related Processes

Fermentation of sugar to ethanol and CO_2 can also be done by Zymomonas mobilis, however the path is slightly different since formation of pyruvate does not happen by glycolysis but instead by the Entner–Doudoroff pathway. Other microorganisms can produce ethanol from sugars by fermentation but often only as a side product. Examples are

- Heterolactic acid fermentation in which Leuconostoc bacterias produce Lactate + Ethanol + CO_2
- Mixed acid fermentation where Escherichia produce Ethanol mixed with Lactate, Acetate, Succinate, Formate, CO_2 and H_2
- 2,3-butanediol fermentation by Enterobacter producing Ethanol, Butanediol, Lactate, Formate, CO_2 and H_2

Effect of Oxygen

Fermentation does not require oxygen. If oxygen is present, some species of yeast (e.g., *Kluyveromyces lactis* or *Kluyveromyces lipolytica*) will oxidize pyruvate completely to carbon dioxide and water. This process is called cellular respiration. But these species of yeast will produce ethanol only in an anaerobic environment (not cellular respiration).

However, many yeasts such as the commonly used baker's yeast *Saccharomyces cerevisiae*, or fission yeast *Schizosaccharomyces pombe*, prefer fermentation to respiration. These yeasts will produce ethanol even under aerobic conditions, if they are provided with the right kind of nutrition. During batch fermentation, the rate of ethanol production per milligram of cell protein is maximal for a brief period early in this process and declines progressively as ethanol accumulates in the surrounding broth. Studies demonstrate that the removal of this accumulated ethanol does not immediately restore fermentative activity, and they provide evidence that the decline in metabolic rate is due to physiological changes (including possible ethanol damage) rather than to the presence of ethanol. Several potential causes for the decline in fermentative activity have been investigated. Viability remained at or above 90%, internal pH remained near neutrality, and the specific activities of the glycolytic and alcohologenic enzymes (measured in vitro) remained high throughout batch fermentation. None of these factors appears to be causally related to the fall in fermentative activity during batch fermentation.

Bread Baking

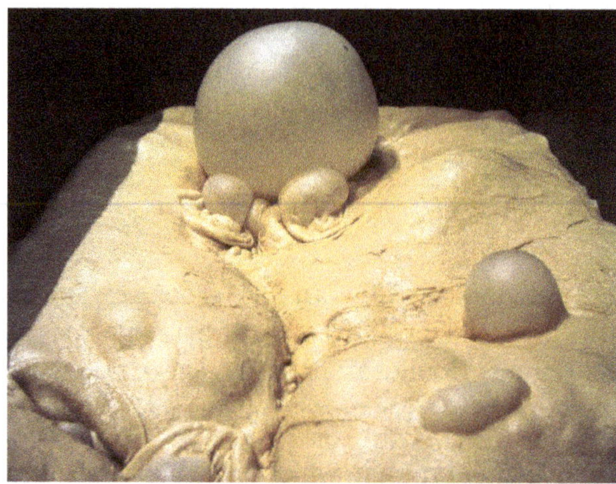

The formation of carbon dioxide — a byproduct of ethanol fermentation — causes bread to rise.

Ethanol fermentation causes bread dough to rise. Yeast organisms consume sugars in the dough and produce ethanol and carbon dioxide as waste products. The carbon dioxide forms bubbles in the dough, expanding it into something of a foam. Less than 2% ethanol remains after baking.

Alcoholic Beverages

All ethanol contained in alcoholic beverages (including ethanol produced by carbonic maceration) is produced by means of fermentation induced by yeast.

- Wine is produced by fermentation of the natural sugars present in grapes; cider and perry are produced by similar fermentation of natural sugar in apples and pears, respectively; and other fruit wines are produced from the fermentation of the sugars in any other kinds of fruit. Brandy and eaux de vie (e.g. slivovitz) are produced by distillation of these fruit-fermented beverages.

Primary fermentation cellar, Budweiser Brewery, Fort Collins, Colorado

- Mead is produced by fermentation of the natural sugars present in honey.
- Beer, whiskey, and vodka are produced by fermentation of grain starches that have been converted to sugar by the enzyme amylase, which is present in grain kernels that have been malted (i.e. germinated). Other sources of starch (e.g. potatoes and unmalted grain) may be added to the mixture, as the amylase will act on those starches as well. Whiskey and vodka are also distilled; gin and related beverages are produced by the addition of flavoring agents to a vodka-like feedstock during distillation.
- Rice wines (including sake) are produced by the fermentation of grain starches converted to sugar by the mold *Aspergillus oryzae*. *Baijiu*, *soju*, and *shōchū* are distilled from the product of such fermentation.
- Rum and some other beverages are produced by fermentation and distillation of sugarcane. Rum is usually produced from the sugarcane product molasses.

In all cases, fermentation must take place in a vessel that allows carbon dioxide to escape but prevents outside air from coming in. This is because exposure to oxygen would prevent the formation of ethanol, while a buildup of carbon dioxide creates a risk the vessel will rupture or fail catastrophically, causing injury and property damage.

Feedstocks for Fuel Production

Yeast fermentation of various carbohydrate products is also used to produce the ethanol that is added to gasoline.

The dominant ethanol feedstock in warmer regions is sugarcane. In temperate regions, corn or sugar beets are used.

In the United States, the main feedstock for the production of ethanol is currently corn. Approximately 2.8 gallons of ethanol are produced from one bushel of corn (0.42 liter per kilogram). While much of the corn turns into ethanol, some of the corn also yields by-products such as DDGS (distillers dried grains with solubles) that can be used as feed for livestock. A bushel of corn produces about 18 pounds of DDGS (320 kilograms of DDGS per metric ton of maize). Although most of the fermentation plants have been built in corn-producing regions, sorghum is also an important feedstock for ethanol production in the Plains states. Pearl millet is showing promise as an ethanol feedstock for the southeastern U.S. and the potential of duckweed is being studied.

In some parts of Europe, particularly France and Italy, grapes have become a *de facto* feedstock for fuel ethanol by the distillation of surplus wine. In Japan, it has been proposed to use rice normally made into sake as an ethanol source.

Cassava as Ethanol Feedstock

Ethanol can be made from mineral oil or from sugars or starches. Starches are cheapest. The starchy crop with highest energy content per acre is cassava, which grows in tropical countries.

Thailand already had a large cassava industry in the 1990s, for use as cattle feed and as a cheap admixture to wheat flour. Nigeria and Ghana are already establishing cassava-to-ethanol plants. Production of ethanol from cassava is currently economically feasible when crude oil prices are above US$120 per barrel.

New varieties of cassava are being developed, so the future situation remains uncertain. Currently, cassava can yield between 25-40 tonnes per hectare (with irrigation and fertilizer), and from a tonne of cassava roots, circa 200 liters of ethanol can be produced (assuming cassava with 22% starch content). A liter of ethanol contains circa 21.46 MJ of energy. The overall energy efficiency of cassava-root to ethanol conversion is circa 32%.

The yeast used for processing cassava is *Endomycopsis fibuligera*, sometimes used together with bacterium *Zymomonas mobilis*.

Byproducts of Fermentation

Ethanol fermentation produces unharvested byproducts such as heat, carbon dioxide, food for livestock, and water.

References

- Zehnder, Alexander J. B. (1978). “Ecology of methane formation”. In Mitchell, Ralph. Water pollution microbiology 2. New York: Wiley. pp. 349–376. ISBN 978-0-471-01902-2.

- Anaerobic Digestion Initiative Advisory Committee (ADIAC). "Feedstock". Archived from the original on 13 December 2011.
- Gupta, Sujata (2010-11-06). "Biogas comes in from the cold". New Scientist. London: Sunita Harrington. p. 14. Retrieved 2011-02-04.
- National Non-Food Crops Centre. "NNFCC Renewable Fuels and Energy Factsheet: Anaerobic Digestion", Retrieved on 2011-11-22
- Shah, Dhruti (5 October 2010). "Oxfordshire town sees human waste used to heat homes". BBC NEWS. Archived from the original on 5 October 2010. Retrieved 5 October 2010.

4

Pasteur Effect: An Integrated Study

Yeast is a single cell eukaryote that can be classified as facultative anaerobic organism as it undergoes respiration in the presence of oxygen and ferments when presented with oxygen-deprived conditions. Pyruvic acid, which is the product of glycolysis in the absence of oxygen is broken down into lactate in animals and ethyl alcohol in plants and microbes. This chapter discusses the important molecules that contribute to fermentation like pyruvic acid, acetyl-CoA and the citric acid cycle or the Krebs cycle.

Pasteur Effect

The Pasteur effect is an inhibiting effect of oxygen on the fermentation process.

Discovery

The effect was discovered in 1857 by Louis Pasteur, who showed that aerating yeasted broth causes yeast cell growth to increase, while conversely, fermentation rate decreases.

Explanation

The effect can be easily explained; as the yeast being facultative anaerobes can produce energy using two different metabolic pathways. While the oxygen concentration is low, the product of glycolysis, pyruvate, is turned into ethanol and carbon dioxide, and the energy production efficiency is low (2 moles of ATP per mole of glucose). If the oxygen concentration grows, pyruvate is converted to acetyl CoA that can be used in the citric acid cycle, which increases the efficiency to 32 moles of ATP per mole of glucose. Therefore, about 16 times as much glucose must be consumed anaerobically as aerobically to yield the same amount of ATP.

Under anaerobic conditions, the rate of glucose metabolism is faster, but the amount of ATP produced (as already mentioned) is smaller. When exposed to aerobic conditions, the ATP and Citrate production increases and the rate of glycolysis slows, because the ATP and citrate produced act as allosteric inhibitors for phosphofructokinase 1, the third enzyme in the glycolysis pathway. The Pasteur effect will only occur if glucose concentrations are low (<2 g/L) and if other nutrients, mostly nitrogen, are limited.

So, from the standpoint of ATP production, it is advantageous for yeast to utilize the Krebs Cycle in the presence of oxygen, as more ATP is produced from less glucose;

however, Boulton et. al. (1996) have maintained that yeast will follow the anaerobic, rather than aerobic, fermentative pathway if glucose is not limited, in that respiration, although capable of producing more ATP than glycolysis per molecule of glucose, also requires more energy in terms of enzymatic and mitochondrial requirements.

Practical Implications

The processes used in alcohol production are commonly maintained in a low oxygen condition, under a blanket of carbon dioxide, while breeding yeast for biomass is done in aerobic conditions, the broth being aerated.

Yeast

Yeasts are eukaryotic, single-celled microorganisms classified as members of the fungus kingdom. The yeast lineage originated hundreds of millions of years ago, and 1,500 species are currently identified. They are estimated to constitute 1% of all described fungal species. Yeasts are unicellular organisms which evolved from multicellular ancestors, with some species having the ability to develop multicellular characteristics by forming strings of connected budding cells known as pseudohyphae or false hyphae. Yeast sizes vary greatly, depending on species and environment, typically measuring 3–4 µm in diameter, although some yeasts can grow to 40 µm in size. Most yeasts reproduce asexually by mitosis, and many do so by the asymmetric division process known as budding.

Yeasts, with their single-celled growth habit, can be contrasted with molds, which grow hyphae. Fungal species that can take both forms (depending on temperature or other conditions) are called dimorphic fungi ("dimorphic" means "having two forms").

By fermentation, the yeast species *Saccharomyces cerevisiae* converts carbohydrates to carbon dioxide and alcohols – for thousands of years the carbon dioxide has been used in baking and the alcohol in alcoholic beverages. It is also a centrally important model organism in modern cell biology research, and is one of the most thoroughly researched eukaryotic microorganisms. Researchers have used it to gather information about the biology of the eukaryotic cell and ultimately human biology. Other species of yeasts, such as *Candida albicans*, are opportunistic pathogens and can cause infections in humans. Yeasts have recently been used to generate electricity in microbial fuel cells, and produce ethanol for the biofuel industry.

Yeasts do not form a single taxonomic or phylogenetic grouping. The term "yeast" is often taken as a synonym for *Saccharomyces cerevisiae*, but the phylogenetic diversity of yeasts is shown by their placement in two separate phyla: the Ascomycota and the Basidiomycota. The budding yeasts ("true yeasts") are classified in the order Saccharomycetales, within the phylum Ascomycota.

History

The word "yeast" comes from Old English *gist*, *gyst*, and from the Indo-European root *yes*-, meaning "boil", "foam", or "bubble". Yeast microbes are probably one of the earliest domesticated organisms. Archaeologists digging in Egyptian ruins found early grinding stones and baking chambers for yeast-raised bread, as well as drawings of 4,000-year-old bakeries and breweries. In 1680, Dutch naturalist Anton van Leeuwenhoek first microscopically observed yeast, but at the time did not consider them to be living organisms, but rather globular structures. Researchers were doubtful whether yeasts were algae or fungi, but in 1837 Theodor Schwann recognized them as fungi.

In 1857, French microbiologist Louis Pasteur proved in the paper "*Mémoire sur la fermentation alcoolique*" that alcoholic fermentation was conducted by living yeasts and not by a chemical catalyst. Pasteur showed that by bubbling oxygen into the yeast broth, cell growth could be increased, but fermentation was inhibited – an observation later called the "Pasteur effect".

By the late 18th century, two yeast strains used in brewing had been identified: *Saccharomyces cerevisiae* (top-fermenting yeast) and *S. carlsbergensis* (bottom-fermenting yeast). *S. cerevisiae* has been sold commercially by the Dutch for bread-making since 1780; while, around 1800, the Germans started producing *S. cerevisiae* in the form of cream. In 1825, a method was developed to remove the liquid so the yeast could be prepared as solid blocks. The industrial production of yeast blocks was enhanced by the introduction of the filter press in 1867. In 1872, Baron Max de Springer developed a manufacturing process to create granulated yeast, a technique that was used until the first World War. In the United States, naturally occurring airborne yeasts were used almost exclusively until commercial yeast was marketed at the Centennial Exposition in 1876 in Philadelphia, where Charles L. Fleischmann exhibited the product and a process to use it, as well as serving the resultant baked bread.

Nutrition and Growth

Yeasts are chemoorganotrophs, as they use organic compounds as a source of energy and do not require sunlight to grow. Carbon is obtained mostly from hexose sugars, such as glucose and fructose, or disaccharides such as sucrose and maltose. Some species can metabolize pentose sugars such as ribose, alcohols, and organic acids. Yeast species either require oxygen for aerobic cellular respiration (obligate aerobes) or are anaerobic, but also have aerobic methods of energy production (facultative anaerobes). Unlike bacteria, no known yeast species grow only anaerobically (obligate anaerobes). Yeasts grow best in a neutral or slightly acidic pH environment.

Yeasts vary in regard to the temperature range in which they grow best. For example, *Leucosporidium frigidum* grows at −2 to 20 °C (28 to 68 °F), *Saccharomyces telluris* at 5 to 35 °C (41 to 95 °F), and *Candida slooffi* at 28 to 45 °C (82 to 113 °F). The cells can survive freezing under certain conditions, with viability decreasing over time.

In general, yeasts are grown in the laboratory on solid growth media or in liquid broths. Common media used for the cultivation of yeasts include potato dextrose agar or potato dextrose broth, Wallerstein Laboratories nutrient agar, yeast peptone dextrose agar, and yeast mould agar or broth. Home brewers who cultivate yeast frequently use dried malt extract and agar as a solid growth medium. The antibiotic cycloheximide is sometimes added to yeast growth media to inhibit the growth of *Saccharomyces* yeasts and select for wild/indigenous yeast species. This will change the yeast process.

The appearance of a white, thready yeast, commonly known as kahm yeast, is often a byproduct of the lactofermentation (or pickling) of certain vegetables, usually the result of exposure to air. Although harmless, it can give pickled vegetables a bad flavor and must be removed regularly during fermentation.

Ecology

Yeasts are very common in the environment, and are often isolated from sugar-rich materials. Examples include naturally occurring yeasts on the skins of fruits and berries (such as grapes, apples, or peaches), and exudates from plants (such as plant saps or cacti). Some yeasts are found in association with soil and insects. The ecological function and biodiversity of yeasts are relatively unknown compared to those of other microorganisms. Yeasts, including *Candida albicans*, *Rhodotorula rubra*, *Torulopsis* and *Trichosporon cutaneum*, have been found living in between people's toes as part of their skin flora. Yeasts are also present in the gut flora of mammals and some insects and even deep-sea environments host an array of yeasts.

An Indian study of seven bee species and 9 plant species found 45 species from 16 genera colonise the nectaries of flowers and honey stomachs of bees. Most were members of the *Candida* genus; the most common species in honey stomachs was *Dekkera intermedia* and in flower nectaries, *Candida blankii*. Yeast colonising nectaries of the stinking hellebore have been found to raise the temperature of the flower, which may aid in attracting pollinators by increasing the evaporation of volatile organic compounds. A black yeast has been recorded as a partner in a complex relationship between ants, their mutualistic fungus, a fungal parasite of the fungus and a bacterium that kills the parasite. The yeast has a negative effect on the bacteria that normally produce antibiotics to kill the parasite, so may affect the ants' health by allowing the parasite to spread.

Certain strains of some species of yeasts produce proteins called yeast killer toxins that allow them to eliminate competing strains. This can cause problems for winemaking but could potentially also be used to advantage by using killer toxin-producing strains to make the wine. Yeast killer toxins may also have medical applications in treating yeast infections.

Reproduction

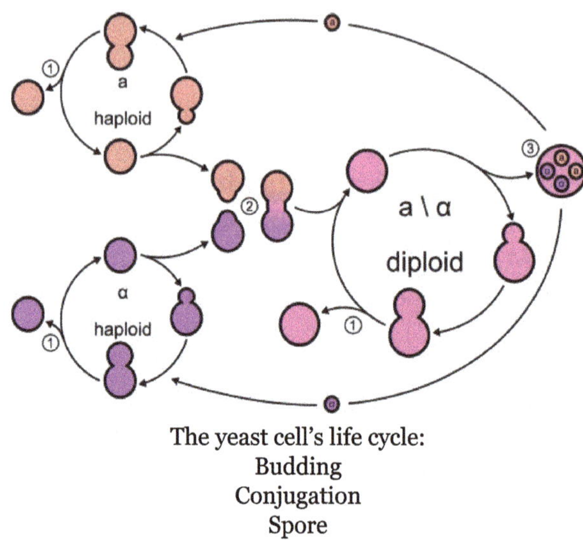

The yeast cell's life cycle:
Budding
Conjugation
Spore

Yeasts, like all fungi, may have asexual and sexual reproductive cycles. The most common mode of vegetative growth in yeast is asexual reproduction by budding. Here, a small bud (also known as a bleb), or daughter cell, is formed on the parent cell. The nucleus of the parent cell splits into a daughter nucleus and migrates into the daughter cell. The bud continues to grow until it separates from the parent cell, forming a new cell. The daughter cell produced during the budding process is generally smaller than the mother cell. Some yeasts, including *Schizosaccharomyces pombe*, reproduce by fission instead of budding, thereby creating two identically sized daughter cells.

In general, under high-stress conditions such as nutrient starvation, haploid cells will die; under the same conditions, however, diploid cells can undergo sporulation, entering sexual reproduction (meiosis) and producing a variety of haploid spores, which can go on to mate (conjugate), reforming the diploid.

The haploid fission yeast *Schizosaccharomyces pombe* is a facultative sexual microorganism that can undergo mating when nutrients are limiting. Exposure of *S. pombe* to hydrogen peroxide, an agent that causes oxidative stress leading to oxidative DNA damage, strongly induces mating and the formation of meiotic spores. The budding yeast *Saccharomyces cerevisiae* reproduces by mitosis as diploid cells when nutrients are abundant, but when starved, this yeast undergoes meiosis to form haploid spores. Haploid cells may then reproduce asexually by mitosis. Katz Ezov et al. presented evidence that in natural *S. cerevisiae* populations clonal reproduction and selfing (in the form of intratetrad mating) predominate. In nature, mating of haploid cells to form diploid cells is most often between members of the same clonal population and out-crossing is uncommon. Analysis of the ancestry of natural *S. cerevisiae* strains led to the conclusion that out-crossing occurs only about once every 50,000 cell divisions. These observations suggest that the possible long-term benefits of outcrossing (e.g. generation of diversity) are likely to be insufficient for generally maintaining sex from one generation to the next. Rather, a short-term benefit, such as recombinational repair during meiosis, may be the key to the maintenance of sex in *S. cerevisiae.*

Some pucciniomycete yeasts, in particular species of *Sporidiobolus* and *Sporobolomyces*, produce aerially dispersed, asexual ballistoconidia.

Uses

The useful physiological properties of yeast have led to their use in the field of biotechnology. Fermentation of sugars by yeast is the oldest and largest application of this technology. Many types of yeasts are used for making many foods: baker's yeast in bread production, brewer's yeast in beer fermentation, and yeast in wine fermentation and for xylitol production. So-called red rice yeast is actually a mold, *Monascus purpureus*. Yeasts include some of the most widely used model organisms for genetics and cell biology.

Alcoholic Beverages

Alcoholic beverages are defined as beverages that contain ethanol (C_2H_5OH). This ethanol is almost always produced by fermentation – the metabolism of carbohydrates by certain species of yeasts under anaerobic or low-oxygen conditions. Beverages such as mead, wine, beer, or distilled spirits all use yeast at some stage of their production. A distilled beverage is a beverage containing ethanol that has been purified by distillation. Carbohydrate-containing plant material is fermented by yeast, producing a dilute solution of ethanol in the process. Spirits such as whiskey and rum are prepared by distilling these dilute solutions of ethanol. Components other than ethanol are collected in the condensate, including water, esters, and other alcohols, which (in addition to that provided by the oak in which it may be aged) account for the flavour of the beverage.

Beer

Yeast ring used by Swedish homebrewers in the 19th century to preserve yeast between brewing sessions.

Brewing yeasts may be classed as "top-cropping" (or "top-fermenting") and "bottom-cropping" (or "bottom-fermenting"). Top-cropping yeasts are so called because they form a foam at the top of the wort during fermentation. An example of a top-cropping yeast is *Saccharomyces cerevisiae*, sometimes called an "ale yeast". Bottom-cropping yeasts are typically used to produce lager-type beers, though they can also produce ale-type beers. These yeasts ferment well at low temperatures. An example of bottom-cropping yeast is *Saccharomyces pastorianus*, formerly known as *S. carlsbergensis*.

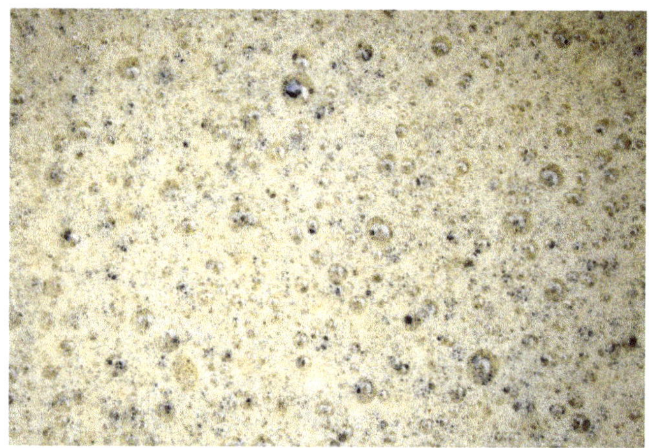

Bubbles of carbon dioxide forming during beer-brewing

Decades ago, taxonomists reclassified *S. carlsbergensis* (uvarum) as a member of *S. cerevisiae*, noting that the only distinct difference between the two is metabolic. Lager strains of *S. cerevisiae* secrete an enzyme called melibiase, allowing them to hydrolyse melibiose, a disaccharide, into more fermentable monosaccharides. Top- and bottom-cropping and cold- and warm-fermenting distinctions are largely generalizations used by laypersons to communicate to the general public.

The most common top-cropping brewer's yeast, *S. cerevisiae*, is the same species as the common baking yeast. Brewer's yeast is also very rich in essential minerals and the B vitamins (except B_{12}). However, baking and brewing yeasts typically belong to different strains, cultivated to favour different characteristics: baking yeast strains are more aggressive, to carbonate dough in the shortest amount of time possible; brewing yeast strains act slower but tend to produce fewer off-flavours and tolerate higher alcohol concentrations (with some strains, up to 22%).

Dekkera/Brettanomyces is a genus of yeast known for its important role in the production of 'lambic' and specialty sour ales, along with the secondary conditioning of a particular Belgian Trappist beer. The taxonomy of the genus *Brettanomyces* has been debated since its early discovery and has seen many reclassifications over the years. Early classification was based on a few species that reproduced asexually (anamorph form) through multipolar budding. Shortly after, the formation of ascospores was observed and the genus *Dekkera*, which reproduces sexually (teleomorph form), was introduced as part of the taxonomy. The current taxonomy includes five species within the genera of *Dekkera/Brettanomyces*. Those are the anamorphs *Brettanomyces bruxellensis*, *Brettanomyces anomalus*, *Brettanomyces custersianus*, *Brettanomyces naardenensis*, and *Brettanomyces nanus*, with teleomorphs existing for the first two species, *Dekkera bruxellensis* and *Dekkera anomala*. The distinction between *Dekkera* and *Brettanomyces* is arguable, with Oelofse et al. (2008) citing Loureiro and Malfeito-Ferreira from 2006 when they affirmed that current molecular DNA detection techniques have uncovered no variance between the anamorph and teleomorph states. Over the past decade, *Brettanomyces* spp. have seen an increasing use in the craft-brewing sector of the industry, with a handful of breweries having produced beers that were primarily fermented with pure cultures of *Brettanomyces* spp. This has occurred out of experimentation, as very little information exists regarding pure culture fermentative capabilities and the aromatic compounds produced by various strains. *Dekkera/Brettanomyces* spp. have been the subjects of numerous studies conducted over the past century, although a majority of the recent research has focused on enhancing the knowledge of the wine industry. Recent research on eight *Brettanomyces* strains available in the brewing industry focused on strain-specific fermentations and identified the major compounds produced during pure culture anaerobic fermentation in wort.

Wine

Yeast is used in winemaking, where it converts the sugars present (glucose and fructose) in grape juice (must) into ethanol. Yeast is normally already present on grape skins. Fermentation can be done with this endogenous "wild yeast", but this procedure gives unpredictable results, which depend upon the exact types of yeast species present. For this reason, a pure yeast culture is usually added to the must; this yeast quickly dominates the fermentation. The wild yeasts are repressed, which ensures a reliable and predictable fermentation.

Most added wine yeasts are strains of *S. cerevisiae*, though not all strains of the species are suitable. Different *S. cerevisiae* yeast strains have differing physiological and fermentative properties, therefore the actual strain of yeast selected can have a direct impact on the finished wine. Significant research has been undertaken into the development of novel wine yeast strains that produce atypical flavour profiles or increased complexity in wines.

The growth of some yeasts, such as *Zygosaccharomyces* and *Brettanomyces*, in wine can result in wine faults and subsequent spoilage. *Brettanomyces* produces an array of metabolites when growing in wine, some of which are volatile phenolic compounds. Together, these compounds are often referred to as "*Brettanomyces* character", and are often described as "antiseptic" or "barnyard" type aromas. *Brettanomyces* is a significant contributor to wine faults within the wine industry.

Researchers from the University of British Columbia, Canada, have found a new strain of yeast that has reduced amines. The amines in red wine and Chardonnay produce off-flavors and cause headaches and hypertension in some people. About 30% of people are sensitive to biogenic amines, such as histamines.

Baking

Yeast, the most common one being *S. cerevisiae*, is used in baking as a leavening agent, where it converts the food/fermentable sugars present in dough into the gas carbon dioxide. This causes the dough to expand or rise as gas forms pockets or bubbles. When the dough is baked, the yeast dies and the air pockets "set", giving the baked product a soft and spongy texture. The use of potatoes, water from potato boiling, eggs, or sugar in a bread dough accelerates the growth of yeasts. Most yeasts used in baking are of the same species common in alcoholic fermentation. In addition, *Saccharomyces exiguus* (also known as *S. minor*), a wild yeast found on plants, fruits, and grains, is occasionally used for baking. In breadmaking, the yeast initially respires aerobically, producing carbon dioxide and water. When the oxygen is depleted, fermentation begins, producing ethanol as a waste product; however, this evaporates during baking.

A block of compressed fresh yeast

It is not known when yeast was first used to bake bread. The first records that show this use came from Ancient Egypt. Researchers speculate a mixture of flour meal and water was left longer than usual on a warm day and the yeasts that occur in natural contaminants of the flour caused it to ferment before baking. The resulting bread would have been lighter and tastier than the normal flat, hard cake.

Today, there are several retailers of baker's yeast; one of the earlier developments in North America is Fleischmann's Yeast, in 1868. During World War II, Fleischmann's developed a granulated active dry yeast which did not require refrigeration, had a longer shelf life than fresh yeast, and rose twice as fast. Baker's yeast is also sold as a fresh yeast compressed into a square "cake". This form perishes quickly, so must be used soon after production. A weak solution of water and sugar can be used to determine whether yeast is expired. In the solution, active yeast will foam and bubble as it ferments the sugar into ethanol and carbon dioxide. Some recipes refer to this as proofing the yeast, as it "proves" (tests) the viability of the yeast before the other ingredients are added. When a sourdough starter is used, flour and water are added instead of sugar; this is referred to as proofing the sponge.

Active dried yeast, a granulated form in which yeast is commercially sold

When yeast is used for making bread, it is mixed with flour, salt, and warm water or milk. The dough is kneaded until it is smooth, and then left to rise, sometimes until it has doubled in size. The dough is then shaped into loaves. Some bread doughs are knocked back after one rising and left to rise again (this is called dough proofing) and then baked. A longer rising time gives a better flavour, but the yeast can fail to raise the bread in the final stages if it is left for too long initially.

Bioremediation

Some yeasts can find potential application in the field of bioremediation. One such yeast, *Yarrowia lipolytica*, is known to degrade palm oil mill effluent, TNT (an explosive material), and other hydrocarbons, such as alkanes, fatty acids, fats and oils. It can also tolerate high concentrations of salt and heavy metals, and is being investigated for its potential as a heavy metal biosorbent. *Saccharomyces cerevisiae* has potential to bioremediate toxic pollutants like arsenic from industrial effluent. Bronze statues are known to be degraded by certain species of yeast. Different yeasts from Brazilian gold mines bioaccumulate free and complexed silver ions.

Industrial Ethanol Production

The ability of yeast to convert sugar into ethanol has been harnessed by the biotechnology industry to produce ethanol fuel. The process starts by milling a feedstock, such as sugar cane, field corn, or other cereal grains, and then adding dilute sulfuric acid, or fungal alpha amylase enzymes, to break down the starches into complex sugars. A glucoamylase is then added to break the complex sugars down into simple sugars. After this, yeasts are added to convert the simple sugars to ethanol, which is then distilled off to obtain ethanol up to 96% in purity.

Saccharomyces yeasts have been genetically engineered to ferment xylose, one of the major fermentable sugars present in cellulosic biomasses, such as agriculture residues, paper wastes, and wood chips. Such a development means ethanol can be efficiently produced from more inexpensive feedstocks, making cellulosic ethanol fuel a more competitively priced alternative to gasoline fuels.

Nonalcoholic Beverages

A kombucha culture fermenting in a jar

A number of sweet carbonated beverages can be produced by the same methods as beer, except the fermentation is stopped sooner, producing carbon dioxide, but only trace amounts of alcohol, leaving a significant amount of residual sugar in the drink.

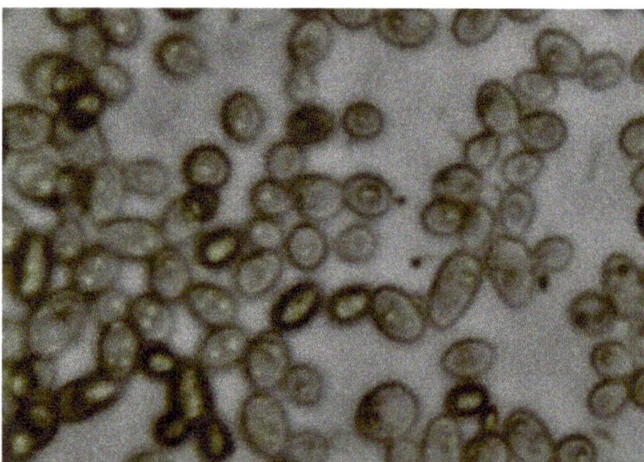

Yeast and bacteria in kombucha at 400X

- Root beer, originally made by Native Americans, commercialized in the United States by Charles Elmer Hires and especially popular during Prohibition

- Kvass, a fermented drink made from rye, popular in Eastern Europe. It has a recognizable, but low alcoholic content.
- Kombucha, a fermented sweetened tea. Yeast in symbiosis with acetic acid bacteria is used in its preparation. Species of yeasts found in the tea can vary, and may include: *Brettanomyces bruxellensis, Candida stellata, Schizosaccharomyces pombe, Torulaspora delbrueckii* and *Zygosaccharomyces bailii.* Also popular in Eastern Europe and some former Soviet republics under the name *chajnyj grib* (Russian: Чайный гриб), which means "tea mushroom".
- Kefir and kumis are made by fermenting milk with yeast and bacteria.
- Mauby (Spanish: *mabí*), made by fermenting sugar with the wild yeasts naturally present on the bark of the *Colubrina elliptica* tree, popular in the Caribbean

Nutritional Supplements

Yeast is used in nutritional supplements,especially those marketed to vegans. It is often referred to as "nutritional yeast" when sold as a dietary supplement. Nutritional yeast is a deactivated yeast, usually *S. cerevisiae*. It is an excellent source of protein and vitamins, especially the B-complex vitamins, as well as other minerals and cofactors required for growth. It is also naturally low in fat and sodium. Contrary to some claims, it contains little or no vitamin B_{12}. Some brands of nutritional yeast, though not all, are fortified with vitamin B_{12}, which is produced separately by bacteria.

In 1920, the Fleischmann Yeast Company began to promote yeast cakes in a successful "Yeast for Health" campaign. They initially emphasized yeast as a source of vitamins, good for skin and digestion. Their later advertising claimed a much broader range of health benefits, and was censured as misleading by the Federal Trade Commission. The fad for yeast cakes lasted until the late 1930s.

Nutritional yeast has a nutty, cheesy flavor and is often used as an ingredient in cheese substitutes. Another popular use is as a topping for popcorn. It can also be used in mashed and fried potatoes, as well as in scrambled eggs. It comes in the form of flakes, or as a yellow powder similar in texture to cornmeal. In Australia, it is sometimes sold as "savoury yeast flakes". Though "nutritional yeast" usually refers to commercial products, inadequately fed prisoners have used "home-grown" yeast to prevent vitamin deficiency.

Probiotics

Some probiotic supplements use the yeast *S. boulardii* to maintain and restore the natural flora in the gastrointestinal tract. *S. boulardii* has been shown to reduce the symptoms of acute diarrhea, reduce the chance of infection by *Clostridium difficile* (often identified simply as C. difficile or C. diff), reduce bowel movements in diarrhea-predominant IBS patients, and reduce the incidence of antibiotic-, traveler's-, and HIV/AIDS-associated diarrheas.

Aquarium Hobby

Yeast is often used by aquarium hobbyists to generate carbon dioxide (CO_2) to nourish plants in planted aquaria. CO_2 levels from yeast are more difficult to regulate than those from pressurized CO_2 systems. However, the low cost of yeast makes it a widely used alternative.

Yeast Extract

Marmite and Vegemite, products made from yeast extract

Yeast extract is the common name for various forms of processed yeast products that are used as food additives or flavours. They are often used in the same way that monosodium glutamate (MSG) is used and, like MSG, often contain free glutamic acid. The general method for making yeast extract for food products such as Vegemite and Marmite on a commercial scale is to add salt to a suspension of yeast, making the solution hypertonic, which leads to the cells' shrivelling up. This triggers autolysis, wherein the yeast's digestive enzymes break their own proteins down into simpler compounds, a process of self-destruction. The dying yeast cells are then heated to complete their breakdown, after which the husks (yeast with thick cell walls that would give poor texture) are separated. Yeast autolysates are used in Vegemite and Promite (Australia); Marmite, Bovril and Oxo (the United Kingdom, Republic of Ireland and South Africa); and Cenovis (Switzerland).

Marmite and Vegemite have a distinctive dark colour

Scientific Research

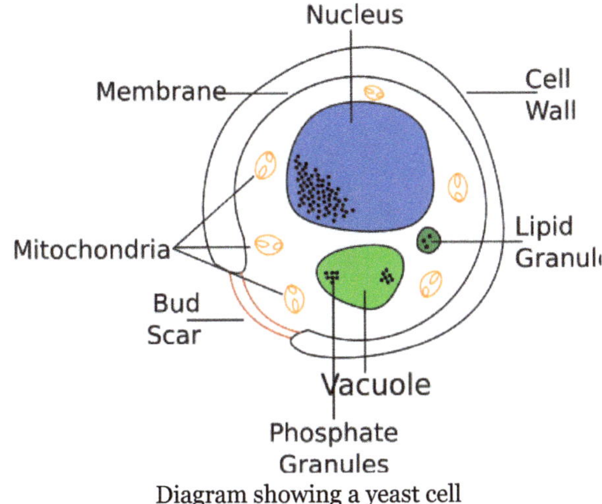

Diagram showing a yeast cell

Several yeasts, in particular *S. cerevisiae*, have been widely used in genetics and cell biology, largely because *S. cerevisiae* is a simple eukaryotic cell, serving as a model for all eukaryotes, including humans, for the study of fundamental cellular processes such as the cell cycle, DNA replication, recombination, cell division, and metabolism. Also, yeasts are easily manipulated and cultured in the laboratory, which has allowed for the development of powerful standard techniques, such as yeast two-hybrid, synthetic genetic array analysis, and tetrad analysis. Many proteins important in human biology were first discovered by studying their homologues in yeast; these proteins include cell cycle proteins, signaling proteins, and protein-processing enzymes.

On 24 April 1996, *S. cerevisiae* was announced to be the first eukaryote to have its genome, consisting of 12 million base pairs, fully sequenced as part of the Genome Project. At the time, it was the most complex organism to have its full genome sequenced, and the work seven years and the involvement of more than 100 laboratories to accomplish. The second yeast species to have its genome sequenced was *Schizosaccharomyces pombe*, which was completed in 2002. It was the sixth eukaryotic genome sequenced and consists of 13.8 million base pairs. As of 2014, over 50 yeast species have had their genomes sequenced and published.

Genetically Engineered Biofactories

Various yeast species have been genetically engineered to efficiently produce various drugs, a technique called metabolic engineering. *S. cerevisiae* is easy to genetically engineer; its physiology, metabolism and genetics are well known, and it is amenable for use in harsh industrial conditions. A wide variety of chemical in different classes can be produced by engineered yeast, including phenolics, isoprenoids, alkaloids, and polyketides. About 20% of biopharmaceuticals are produced in *S. cerevisiae*, including insulin, vaccines for hepatitis, and human serum albumin.

Pathogenic Yeasts

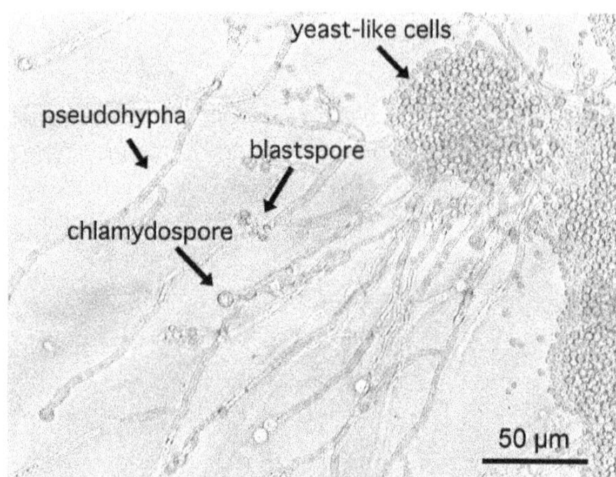

A photomicrograph of Candida albicans showing hyphal outgrowth and other morphological characteristics

Some species of yeast are opportunistic pathogens that can cause infection in people with compromised immune systems. *Cryptococcus neoformans* and *Cryptococcus gattii* are significant pathogens of immunocompromised people. They are the species primarily responsible for cryptococcosis, a fungal disease that occurs in about one million HIV/AIDS patients, causing over 600,000 deaths annually. The cells of these yeast are surrounded by a rigid polysaccharide capsule, which helps to prevent them from being recognised and engulfed by white blood cells in the human body.

Yeasts of the *Candida* genus, another group of opportunistic pathogens, cause oral and vaginal infections in humans, known as candidiasis. *Candida* is commonly found as a commensal yeast in the mucous membranes of humans and other warm-blooded animals. However, sometimes these same strains can become pathogenic. The yeast cells sprout a hyphal outgrowth, which locally penetrates the mucosal membrane, causing irritation and shedding of the tissues. The pathogenic yeasts of candidiasis in probable descending order of virulence for humans are: *C. albicans*, *C. tropicalis*, *C. stellatoidea*, *C. glabrata*, *C. krusei*, *C. parapsilosis*, *C. guilliermondii*, *C. viswanathii*, *C. lusitaniae*, and *Rhodotorula mucilaginosa*. *Candida glabrata* is the second most common *Candida* pathogen after *C. albicans*, causing infections of the urogenital tract, and of the bloodstream (candidemia).

Food Spoilage

Yeasts are able to grow in foods with a low pH (5.0 or lower) and in the presence of sugars, organic acids, and other easily metabolized carbon sources. During their growth, yeasts metabolize some food components and produce metabolic end products. This causes the physical, chemical, and sensible properties of a food to change, and the food is spoiled. The growth of yeast within food products is often seen on their surfaces, as in cheeses or meats, or by the fermentation of sugars in beverages, such as juices, and semiliquid products, such as syrups and jams. The yeast of the *Zygosaccharomyces* genus have had a long history as spoilage yeasts within the food industry. This is mainly because these species can grow in the presence of high sucrose, ethanol, acetic acid, sorbic acid, benzoic acid, and sulphur dioxide concentrations, representing some of the commonly used food preservation methods. Methylene blue is used to test for the presence of live yeast cells. In oenology, the major spoilage yeast is *Brettanomyces bruxellensis*.

Facultative Anaerobic Organism

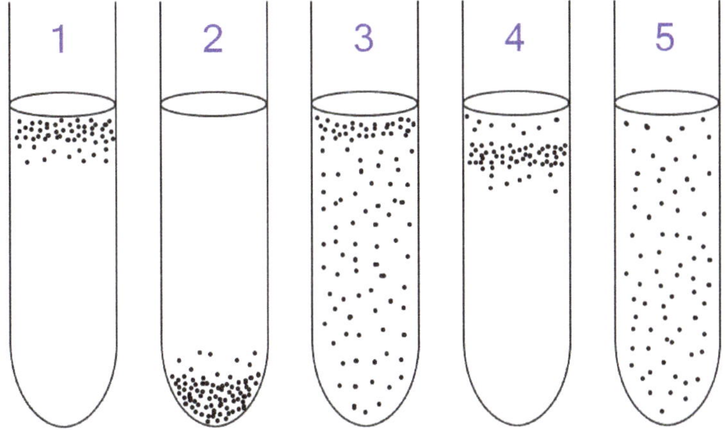

Aerobic and anaerobic bacteria can be identified by growing them in test tubes of thioglycollate broth:1: Obligate aerobes need oxygen because they cannot ferment or respire anaerobically. They gather at the top of the tube where the oxygen concentration is highest.2: Obligate anaerobes are poisoned by oxygen, so they gather at the bottom of the tube where the oxygen concentration is lowest.3: Facultative anaerobes can grow with or without oxygen because they can metabolise energy aerobically or anaerobically. They gather mostly at the top because aerobic respiration generates more ATP than either fermentation or anaerobic respiration.4: Microaerophiles need oxygen because they cannot ferment or respire anaerobically. However, they are poisoned by high concentrations of oxygen. They gather in the upper part of the test tube but not the very top.5: Aerotolerant organisms do not require oxygen as they metabolise energy anaerobically. Unlike obligate anaerobes however, they are not poisoned by oxygen. They can be found evenly spread throughout the test tube.

A facultative anaerobe is an organism that makes ATP by aerobic respiration if oxygen is present, but is capable of switching to fermentation or anaerobic respiration if oxygen is absent. An obligate aerobe, by contrast, cannot make ATP in the absence of oxygen, and obligate anaerobes die in the presence of oxygen.

Some examples of facultatively anaerobic bacteria are *Staphylococcus* spp., *Streptococcus* spp., *Escherichia coli*, *Listeria* spp. and *Shewanella oneidensis*. Certain eukaryotes are also facultative anaerobes, including fungi such as *Saccharomyces cerevisiae* and many aquatic invertebrates such as Nereid (worm) polychaetes.

Various Fermentation Food

Staphylococcus

Staphylococcus is a genus of Gram-positive bacteria. Under the microscope, they appear round (cocci), and form in grape-like clusters.

The *Staphylococcus* genus includes at least 40 species. Of these, nine have two subspecies, one has three subspecies, and one has four subspecies. Most are harmless and reside normally on the skin and mucous membranes of humans and other organisms. Found worldwide, they are a small component of soil microbial flora.

Taxonomy

The taxonomy is based on 16s rRNA sequences, and most of the staphylococcal species fall into 11 clusters:

1. *S. aureus* group – *S. aureus, S. simiae*
2. *S. auricularis* group – *S. auricularis*
3. *S. carnosus* group – *S. carnosus, S. condimenti, S. massiliensis, S. piscifermentans, S. simulans*
4. *S. epidermidis* group – *S. capitis, S. caprae, S. epidermidis, S. saccharolyticus*
5. *S. haemolyticus* group – *S. devriesei, S. haemolyticus, S. hominis*
6. *S. hyicus-intermedius* group – *S. agnetis, S. chromogenes, S. felis, S. delphini, S. hyicus, S. intermedius, S. lutrae, S. microti, S. muscae, S. pseudintermedius, S. rostri, S. schleiferi*
7. *S. lugdunensis* group – *S. lugdunensis*
8. *S. saprophyticus* group – *S. arlettae, S. cohnii, S. equorum, S. gallinarum, S. kloosii, S. leei, S. nepalensis, S. saprophyticus, S. succinus, S. xylosus*
9. *S. sciuri* group – *S. fleurettii, S. lentus, S. sciuri, S. stepanovicii, S. vitulinus*
10. *S. simulans* group – *S. simulans*
11. *S. warneri* group – *S. pasteuri, S. warneri*

A 12th group – that of *S. caseolyticus* – has now been moved to a new genus, *Macrococcus*, the species of which are currently the closest known relatives of *Staphylococcus*.

Subspecies

S. aureus subsp. *aureus*

S. aureus subsp. *anaerobius*

S. capitis subsp. *capitis*

S. capitis subsp. *urealyticus*

S. carnosus subsp. *carnosus*

S. carnosus subsp. *utilis*

S. cohnii subsp. *cohnii*

S. cohnii subsp. *urealyticus*

S. equorum subsp. *equorum*

S. equorum subsp. *linens*

S. hominis subsp. *hominis*

S. hominis subsp. *novobiosepticus*

S petrasii subsp. *croceilyticus*

S petrasii subsp. *jettensis*

S petrasii subsp. *petrasii*

S petrasii subsp. *pragensis*

S. saprophyticus subsp. *bovis*

S. saprophyticus subsp. *saprophyticus*

S. schleiferi subsp. *coagulans*

S. schleiferi subsp. *schleiferi*

S. sciuri subsp. *carnaticus*

S. sciuri subsp. *rodentium*

S. sciuri subsp. *sciuri*

S. succinus subsp. *casei*

S. succinus subsp. *succinus*

Notes

As with all generic names in binomial nomenclature, *Staphylococcus* is capitalized when used alone or with a specific species. Also, the abbreviations *Staph* and *S.* when used with a species (*S. aureus*) are correctly italicized and capitalized (though often errors in this are seen in popular literature). However, *Staphylococcus* is not capitalized or italicized when used in adjectival forms, as in a staphylococcal infection, or as the informal plural (staphylococci).

The *S. saprophyticus* and *S. sciuri* groups are generally novobiocin-resistant, as is *S. hominis* subsp. *novobiosepticus*.

Members of the *S. sciuri* group are oxidase-positive due to their possession of the enzyme cytochrome c oxidase. This group is the only clade within the staphylococci to possess this gene.

The *S. sciuri* group appears to be the closest relations to the genus *Macrococcus*.

S. pulvereri has been shown to be a junior synonym of *S. vitulinus*.

Within these clades, the *S. haemolyticus* and *S. simulans* groups appear to be related, as do the *S. aureus* and *S. epidermidis* groups.

S. lugdunensis appears to be related to the *S. haemolyticus* group.

S. petrasii may be related to *S. haemolyticus*, but this needs to be confirmed.

The taxonomic position of *S. lyticans*, *S. pettenkoferi*, *S. petrasii*, and *S. pseudolugdunensis* has yet to be clarified. The published descriptions of these species do not appear to have been validly published.

Biochemical Identification

Assignment of a strain to the genus *Staphylococcus* requires it to be a Gram-positive coccus that forms clusters, produces catalase, has an appropriate cell wall structure (including peptidoglycan type and teichoic acid presence) and G + C content of DNA in a range of 30–40 mol%.

Staphylococcus species can be differentiated from other aerobic and facultative anaerobic, Gram-positive cocci by several simple tests. *Staphylococcus* species are facultative anaerobes (capable of growth both aerobically and anaerobically). All species grow in the presence of bile salts.

All species were once thought to be coagulase-positive, but this has since been disproven.

Growth can also occur in a 6.5% NaCl solution. On Baird Parker medium, *Staphylococcus* species grow fermentatively, except for *S. saprophyticus*, which grows oxidatively. *Staphylococcus* species are resistant to bacitracin (0.04 U disc: resistance = < 10 mm zone of inhibition) and susceptible to furazolidone (100 μg disc: resistance = < 15 mm zone of inhibition). Further biochemical testing is needed to identify to the species level.

When these bacteria divide, they do so along two axes, so form clumps of bacteria. This is as opposed to streptococci, which divide along one axis, so form chains (*strep* meaning twisted or pliant).

Coagulase Production

One of the most important phenotypical features used in the classification of staphylococci is their ability to produce coagulase, an enzyme that causes blood clot formation.

Six species are currently recognised as being coagulase-positive: *S. aureus*, *S. delphini*, *S. hyicus*, *S. intermedius*, *S. lutrae*,*S. pseudintermedius*, and *S. schleiferi* subsp. *coagulans*. These species belong to two separate groups – the *S. aureus* (*S. aureus* alone) group and the *S. hyicus-intermedius* group (the remaining five). *S. aureus* can also be found as being coagulase-negative.

A seventh species has also been described – *Staphylococcus leei* – from patients with gastritis.

S. aureus is coagulase-positive, meaning it produces coagulase. However, while the majority of *S. aureus* strains are coagulase-positive, some may be atypical in that they do not produce coagulase. *S. aureus* is catalase-positive (meaning that it can produce the enzyme catalase) and able to convert hydrogen peroxide (H_2O_2) to water and oxygen, which makes the catalase test useful to distinguish staphylococci from enterococci and streptococci.

S. pseudintermedius inhabits and sometimes infects the skin of domestic dogs and cats. This organism, too, can carry the genetic material that imparts multiple bacterial resistance. It is rarely implicated in infections in humans, as a zoonosis.

S. epidermidis, a coagulase-negative species, is a commensal of the skin, but can cause severe infections in immunosuppressed patients and those with central venous catheters. *S. saprophyticus*, another coagulase-negative species that is part of the normal vaginal flora, is predominantly implicated in genitourinary tract infections in sexually active young women. In recent years, several other *Staphylococcus* species have been implicated in human infections, notably *S. lugdunensis*, *S. schleiferi*, and *S. caprae*.

Common abbreviations for coagulase-negative staphylococci are CoNS, CNS, or CNST. The American Society for Microbiology abbreviates coagulase-negative staphylococci as "CoNS".

Genomics and Molecular Biology

The first *S. aureus* genomes to be sequenced were those of N315 and Mu50, in 2001. Many more complete *S. aureus* genomes have been submitted to the public databases, making it one of the most extensively sequenced bacteria. The use of genomic data is now widespread and provides a valuable resource for researchers working with *S. aureus*. Whole genome technologies, such as sequencing projects and microarrays, have shown an enormous variety of *S. aureus* strains. Each contains different combinations of surface proteins and different toxins. Relating this information to pathogenic behaviour is one of the major areas of staphylococcal research. The development of molecular typing methods has enabled the tracking of different strains of *S. aureus*. This may lead to better control of outbreak strains. A greater understanding of how the staphylococci evolve, especially due to the acquisition of mobile genetic elements encoding resistance and virulence genes is helping to identify new outbreak strains and may even prevent their emergence.

The widespread incidence of antibiotic resistance across various strains of *S. aureus*, or across different species of *Staphylococcus* has been attributed to horizontal gene transfer of genes encoding antibiotic/metal resistance and virulence. A recent study demonstrated the extent of horizontal gene transfer among *Staphylococcus* to be much greater than previously expected, and encompasses genes with functions beyond antibiotic resistance and virulence, and beyond genes residing within the mobile genetic elements.

Various strains of *Staphylococcus* are available from biological research centres, such as the National Collection of Type Cultures.

Host Range

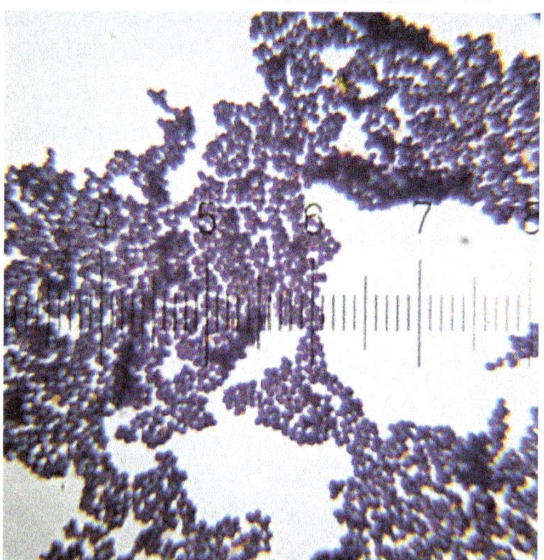

Unknown variety of Staphylococcus, Gram-stained - numbered ticks on the scale are 11 μm apart

Members of the genus *Staphylococcus* frequently colonize the skin and upper respiratory tracts of mammals and birds. Some species specificity has been observed in host range, such that the *Staphylococcus* species observed on some animals appear more rarely on more distantly related host species. Some of the observed host specificity includes:

S. arlattae – chickens, goats

S. aureus - humans

S. auricularis – deer, dogs, humans

S. capitis – humans

S. caprae – goats, humans

S. cohnii – chickens, humans

S. delphini – dolphins

S. devriesei – cattle

S. epidermidis – humans

S. equorum – horses

S. felis – cats

S. fleurettii – goats

S. gallinarum – chickens, goats, pheasants

S. haemolyticus – humans, *Cercocebus*, *Erythrocebus*, *Lemur*, *Macca*, *Microcebus*, *Pan*

S. hyicus – pigs

S. leei – humans

S. lentus – goats, rabbits, sheep

S. lugdunensis – humans, goats

S. lutrae – otters

S. microti – voles (*Microtus arvalis*)

S. nepalensis – goats

S. pasteuri – humans, goats

S. pettenkoferi – humans

S. pseudintermedius – dogs

S. rostri – pigs

S. schleiferi – humans

S. sciuri – humans, dogs, goats

S. simiae – South American squirrel monkeys (*Saimiri sciureus*)

S. simulans – humans

S. warneri – humans, Cercopithecoidea, Pongidae

S. xylosus – humans

Clinical

Staphylococcus can cause a wide variety of diseases in humans and animals through either toxin production or penetration. Staphylococcal toxins are a common cause of food poisoning, for they can be produced by bacteria growing in improperly stored food items. The most common sialadenitis is caused by staphylococci, as bacterial infections.

Streptococcus

Streptococcus is a genus of coccus (spherical) Gram-positive bacteria belonging to the phylum Firmicutes and the order Lactobacillales (lactic acid bacteria). Cell division in this genus occurs along a single axis in these bacteria, thus they grow in chains or pairs, hence the name. (Contrast this with staphylococci, which divide along multiple axes and generate grape-like clusters of cells.)

Most are oxidase-negative and catalase-negative, and many are facultative anaerobes.

In 1984, many bacteria formerly considered *Streptococcus* were separated out into the genera *Enterococcus* and *Lactococcus*. Currently, over 50 species are recognised in this genus.

Pathogenesis and Classification

In addition to streptococcal pharyngitis (strep throat), certain *Streptococcus* species are responsible for many cases of pink eye, meningitis, bacterial pneumonia, endocarditis, erysipelas, and necrotizing fasciitis (the 'flesh-eating' bacterial infections). However, many streptococcal species are not pathogenic, and form part of the commensal human microbiota of the mouth, skin, intestine, and upper respiratory tract. Furthermore, streptococci are a necessary ingredient in producing Emmentaler ("Swiss") cheese.

Species of *Streptococcus* are classified based on their hemolytic properties. Alpha-hemolytic species cause oxidization of iron in hemoglobin molecules within red blood cells, giving it a greenish color on blood agar. Beta-hemolytic species cause complete rupture of red blood cells. On blood agar, this appears as wide areas clear of blood cells surrounding bacterial colonies. Gamma-hemolytic species cause no hemolysis.

Beta-hemolytic streptococci are further classified by Lancefield grouping, a serotype classification (that is, describing specific carbohydrates present on the bacterial cell wall). The 20 described serotypes are named Lancefield groups A to V (excluding I and J).

In the medical setting, the most important groups are the alpha-hemolytic streptococci *S. pneumoniae* and *Streptococcus viridans* group, and the beta-hemolytic streptococci of Lancefield groups A and B (also known as "group A strep" and "group B strep").

Table: Medically Relevant Streptococci (Not All are Alpha hemolytic)

Species	Host	Disease
S. pyogenes	human	pharyngitis
S. agalactiae	human, cattle	neonatal meningitis and sepsis
S. dysgalactiae	human, animals	endocarditis, bacteremia, pneumonia, meningitis, respiratory infections
S. bovis	human, animals	biliary or urinary tract infections, endocarditis
S. anginosus	human, animals	subcutaneous/organ abscesses, meningitis, respiratory infections
S. sanguinis	human	endocarditis, dental caries
S. suis	swine	meningitis
S. mitis	human	endocarditis
S. mutans	human	dental caries
S. pneumoniae	human	pneumonia

Alpha-hemolytic

When alpha hemolysis (α-hemolysis) is present, the agar under the colony is dark and greenish. *Streptococcus pneumoniae* and a group of oral streptococci (*Streptococcus viridans* or viridans streptococci) display alpha hemolysis. This is sometimes called green hemolysis because of the color change in the agar. Other synonymous terms are incomplete hemolysis and partial hemolysis. Alpha hemolysis is caused by hydrogen peroxide produced by the bacterium, oxidizing hemoglobin to green methemoglobin.

Pneumococci

- *S. pneumoniae* (sometimes called pneumococcus), is a leading cause of bacterial pneumonia and occasional etiology of otitis media, sinusitis, meningitis, and peritonitis. Inflammation is thought to be the major cause of how pneumococci cause disease, hence the tendency of diagnoses associated with them to involve inflammation.

The Viridans Group: Alpha-hemolytic

- The viridans streptococci are a large group of commensal bacteria, that are either α-hemolytic, producing a green coloration on blood agar plates (hence the name "viridans", from Latin *vĭrĭdis*, green), or nonhemolytic. They possess no Lancefield antigens.

Beta-hemolytic

Beta hemolysis (β-hemolysis), sometimes called complete hemolysis, is a complete lysis of red cells in the media around and under the colonies: the area appears lightened (yellow) and transparent. Streptolysin, an exotoxin, is the enzyme produced by the bacteria which causes the complete lysis of red blood cells. There are two types of streptolysin: Streptolysin O (SLO) and streptolysin S (SLS). Streptolysin O is an oxygen-sensitive cytotoxin, secreted by most Group A streptococcus (GAS), and interacts with cholesterol in the membrane of eukaryotic cells (mainly red and white blood cells, macrophages, and platelets), and usually results in β-hemolysis under the surface of blood agar. Streptolysin S is an oxygen-stable cytotoxin also produced by most GAS strains which results in clearing on the surface of blood agar. SLS affects immune cells, including polymorphonuclear leukocytes and lymphocytes, and is thought to prevent the host immune system from clearing infection. *Streptococcus pyogenes*, or group A *Streptococcus* (**GAS**), displays beta hemolysis.

Some weakly beta-hemolytic species cause intense beta hemolysis when grown together with a strain of *Staphylococcus*. This is called the CAMP test. *Streptococcus agalactiae* displays this property. *Clostridium perfringens* can be identified presumptively with this test. *Listeria monocytogenes* is also positive on sheep's blood agar.

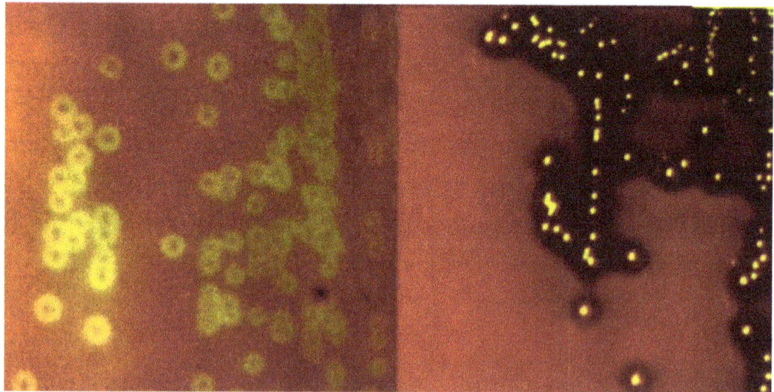

Alpha-hemolytic *S. viridans* (right) and beta-hemolytic *S. pyogenes* (left) streptococci growing on blood agar

Group A

S. pyogenes (GAS) is the causative agent in a wide range of group A streptococcal infections. These infections may be noninvasive or invasive. The noninvasive infections tend to be more common and less severe. The most common of these infections include streptococcal pharyngitis (strep throat) and impetigo. Scarlet fever is also a noninvasive infection, but has not been as common in recent years.

The invasive infections caused by group A β-hemolytic streptococci tend to be more severe and less common. This occurs when the bacterium is able to infect areas where it is not usually found, such as the blood and the organs. The diseases that may be caused include streptococcal toxic shock syndrome, necrotizing fasciitis, pneumonia, and bacteremia. Globally, GAS has been estimated to cause more than 500,000 deaths every year, making it one of the world's leading pathogens.

Additional complications may be caused by GAS, namely acute rheumatic fever and acute glomerulonephritis. Rheumatic fever, a disease that affects the joints, kidneys, and heart valves, is a consequence of untreated strep A infection caused not by the bacterium itself. Rheumatic fever is caused by the antibodies created by the immune system to fight off the infection cross-reacting with other proteins in the body. This "cross-reaction" causes the body to essentially attack itself and leads to the damage above. A similar autoimmune mechanism initiated by Group A beta-hemolytic streptococcal (GABHS) infection is hypothesized to cause pediatric autoimmune neuropsychiatric disorders associated with streptococcal infections (PANDAS), wherein autoimmune antibodies affect the basal ganglia, causing rapid onset of psychiatric, motor, sleep, and other symptoms in pediatric patients.

Group A *Streptococcus* infection is generally diagnosed with a rapid strep test or by culture.

Group B

S. agalactiae, or group B *Streptococcus*, GBS, causes pneumonia and meningitis in neonates and the elderly, with occasional systemic bacteremia. They can also colonize the intestines and the female reproductive tract, increasing the risk for premature rupture of membranes during pregnancy, and transmission of the organism to the infant. The American Congress of Obstetricians and Gynecologists (formerly the American College of Obstetricians and Gynecologists), American Academy of Pediatrics, and the Centers for Disease Control recommend all pregnant women between 35 and 37 weeks gestation to be tested for GBS. Women who test positive should be given prophylactic antibiotics during labor, which will usually prevent transmission to the infant.

The United Kingdom has chosen to adopt a risk factor-based protocol, rather than the culture-based protocol followed in the US. Current guidelines state that if one or more of the following risk factors are present, then women should be treated with *intrapartum* antibiotics:

- Preterm labour (<37 weeks)
- Prolonged rupture of membranes (>18 hours)

- Intrapartum fever (>38C)
- Prior GBS affected infant
- GBS bacteriuria during this pregnancy

This protocol results in treatment of 15–20% of pregnant women and prevention of 65–70% of cases of early onset GBS sepsis.

Group C

This group includes *S. equi*, which causes strangles in horses, and *S. zooepidemicus*—*S. equi* is a clonal descendent or biovar of the ancestral *S. zooepidemicus*—which causes infections in several species of mammals, including cattle and horses. *S. dysgalactiae* is also a member of group C, β-haemolytic streptococci that can cause pharyngitis and other pyogenic infections similar to group A streptococci.

Group D (Enterococci)

Many former group D streptococci have been reclassified and placed in the genus *Enterococcus* (including *E. faecalis*, *E. faecium*, *E. durans*, and *E. avium*). For example, *Streptococcus faecalis* is now *Enterococcus faecalis*. *E. faecalis* is sometimes alpha hemolytic and *E. faecium* is sometimes beta hemolytic.

The remaining nonenterococcal group D strains include *Streptococcus bovis* and *Streptococcus equinus*.

Nonhemolytic streptococci rarely cause illness. However, weakly hemolytic group D beta-hemolytic streptococci and *Listeria monocytogenes* (which is actually a Gram-positive bacillus) should not be confused with nonhemolytic streptococci.

Group F Streptococci

Group F streptococci were first described in 1934 by Long and Bliss amongst the "minute haemolytic streptococci". They are also known as *Streptococcus anginosus* (according to the Lancefield classification system) or as members of the *S. milleri* group (according to the European system).

Group G streptococci

These streptococci are usually, but not exclusively, beta-hemolytic. *Streptococcus dysgalactiae* is the predominant species encountered, particularly in human disease. *S. canis* is an example of a GGS which is typically found on animals, but can cause infection in humans. S. phocae is a GGS subspecies that has been found in marine mammals and marine fish species. In marine mammals it has been mainly associated with meningoencephalitis, septicemia, and endocarditis, but is also associated with many other pathologies. It's environmental reservoir and means of transmission in marine mammals is not well characterized.

Group H Streptococci

Group H streptococci cause infections in medium-sized canines. Group H streptococci rarely cause illness unless a human has direct contact with the mouth of a canine. One of the most common ways this can be spread is human-to-canine, mouth-to-mouth contact. However, the canine may lick the human's hand and infection can be spread, as well.

Molecular Taxonomy and Phylogenetics

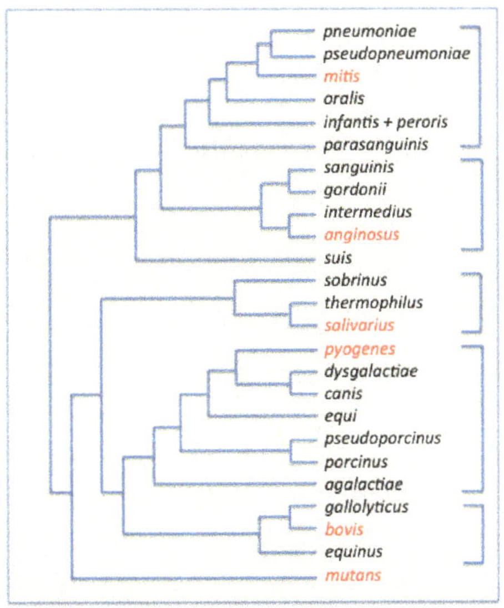

Phylogenetic tree of Streptococcus species, based on data from PATRIC. 16S groups are indicated by brackets and their key members are highlighted in red.

Streptococci have been divided into six groups on the basis of their 16S rDNA sequences: *S. anginosus, S.bovis, S. mitis, S. mutans, S. pyogenes* and *S. salivarius*. The 16S groups have been confirmed by whole genome sequencing (see figure). The important pathogens *S. pneumoniae* and *S. pyogenes* belong to the *S. mitis* and *S. pyogenes* groups, respectively, while the causative agent of dental caries, *Streptococcus mutans*, is basal to the *Streptococcus* group.

Genomics

The genomes of hundreds of species have been sequenced. Most *Streptococcus* genomes are 1.8 to 2.3 Mb in size and encode 1,700 to 2,300 proteins. Some important genomes are listed in the table. The four species shown in the table (*S. pyogenes, S. agalactiae, S. pneumoniae,* and *S. mutans*) have an average pairwise protein sequence identity of about 70%.

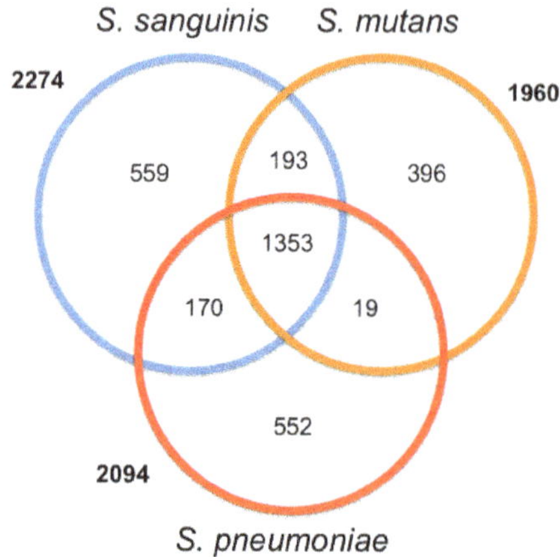

Common and species-specific genes among *Streptococcus sanguinis*, *S. mutans*, and *S. pneumoniae*. Modified after Xu et al. (2007)

feature	*S. pyogenes*	*S. agalactiae*	*S. pneumoniae*	*S. mutans*
base pairs	1,852,442	2,211,488	2,160,837	2,030,921
ORFs	1792	2118	2236	1963
prophages	yes	no	no	no

Bacteriophage

Bacteriophages have been described for many species of *Streptococcus*. 18 prophages have been described in *S. pneumoniae* that range in size from 38 to 41 kb in size, encoding from 42 to 66 genes each. Some of the first *Streptococcus* phages discovered were Dp-1 and ω1. In 1981 the Cp (Complutense phage) family was discovered with Cp-1 as its first member. Dp-1 and Cp-1 infect both *S. pneumoniae* and *S. mitis*. However, the host ranges of most *Streptococcus* phages have not been investigated systematically.

Escherichia

Escherichia is a genus of Gram-negative, non-spore forming, facultatively anaerobic, rod-shaped bacteria from the family Enterobacteriaceae. In those species which are inhabitants of the gastrointestinal tracts of warm-blooded animals, *Escherichia* species provide a portion of the microbially derived vitamin K for their host. er of the species of *Escherichia* are pathogenic. The genus is named after Theodor Escherich, the discoverer of *Escherichia coli*.

Pathogenesis

While many *Escherichia* are commensal gut flora, particular strains of some species, in particular the serotypes of *Escherichia coli* most notably, are human pathogens, and are known as the most common cause of urinary tract infections, significant sources of gastrointestinal disease, ranging from simple diarrhea to dysentery-like conditions, as well as a wide range of other pathogenic states classifiable in general as "colonic escherichiosis." While *E. coli* is responsible for the vast majority of *Escherichia*-related pathogenesis, other members of the genus have also been implicated in human disease. *Escherichia* are associated with the imbalance of microbiota of the lower reproductive tract of women. These species are associated with inflammation.

Pyruvic Acid

Pyruvic acid ($CH_3COCOOH$) is the simplest of the alpha-keto acids, with a carboxylic acid and a ketone functional group. Pyruvate, the conjugate base, CH_3COCOO^-, is a key intermediate in several metabolic pathways.

Pyruvic acid can be made from glucose through glycolysis, converted back to carbohydrates (such as glucose) via gluconeogenesis, or to fatty acids through a reaction with acetyl-CoA. It can also be used to construct the amino acid alanine and can be converted into ethanol or lactic acid via fermentation.

Pyruvic acid supplies energy to cells through the citric acid cycle (also known as the Krebs cycle) when oxygen is present (aerobic respiration), and alternatively ferments to produce lactate when oxygen is lacking (fermentation).

Chemistry

In 1834, Théophile-Jules Pelouze distilled both tartaric acid (L-tartaric acid) and racemic acid (a mix of D- and L-tartaric acid) and isolated pyrotartaric acid (methyl succinic acid) and another acid that Jöns Jacob Berzelius characterized the following year and named pyruvic acid. Pyruvic acid is a colorless liquid with a smell similar to that of acetic acid and is miscible with water. In the laboratory, pyruvic acid may be prepared by heating a mixture of tartaric acid and potassium hydrogen sulfate, by the oxidation of propylene glycol by a strong oxidizer (e.g., potassium permanganate or bleach), or by the hydrolysis of acetyl cyanide, formed by reaction of acetyl chloride with potassium cyanide:

$$CH_3COCl + KCN \rightarrow CH_3COCN + KCl$$

$$CH_3COCN \rightarrow CH_3COCOOH$$

Biochemistry

Pyruvate is an important chemical compound in biochemistry. It is the output of the metabolism of glucose known as glycolysis. One molecule of glucose breaks down into two molecules of pyruvate, which are then used to provide further energy, in one of two ways. Pyruvate is converted into acetyl-coenzyme A, which is the main input for a series of reactions known as the Krebs cycle (also known as the citric acid cycle or tricarboxylic acid cycle). Pyruvate is also converted to oxaloacetate by an anaplerotic reaction, which replenishes Krebs cycle intermediates; also, the oxaloacetate is used for gluconeogenesis. These reactions are named after Hans Adolf Krebs, the biochemist awarded the 1953 Nobel Prize for physiology, jointly with Fritz Lipmann, for research into metabolic processes. The cycle is also known as the citric acid cycle or tricarboxylic acid cycle, because citric acid is one of the intermediate compounds formed during the reactions.

If insufficient oxygen is available, the acid is broken down anaerobically, creating lactate in animals and ethanol in plants and microorganisms. Pyruvate from glycolysis is converted by fermentation to lactate using the enzyme lactate dehydrogenase and the coenzyme NADH in lactate fermentation, or to acetaldehyde and then to ethanol in alcoholic fermentation.

Pyruvate is a key intersection in the network of metabolic pathways. Pyruvate can be converted into carbohydrates via gluconeogenesis, to fatty acids or energy through acetyl-CoA, to the amino acid alanine, and to ethanol. Therefore, it unites several key metabolic processes.

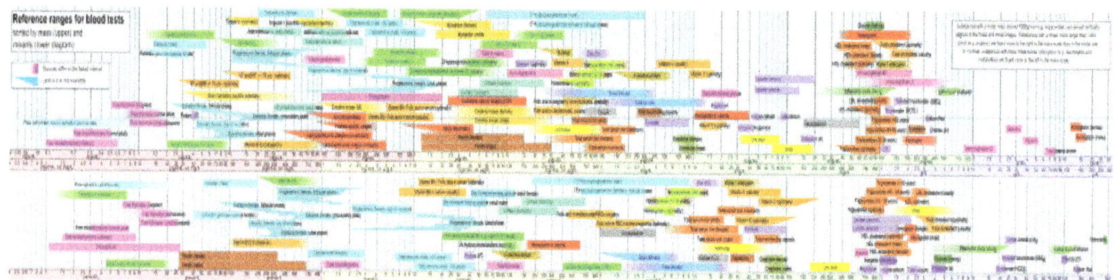

Reference ranges for blood tests, comparing blood content of pyruvate (shown in violet near middle) with other constituents.

Decarboxylation to Acetyl CoA

Pyruvate decarboxylation by the pyruvate dehydrogenase complex produces acetyl-CoA.

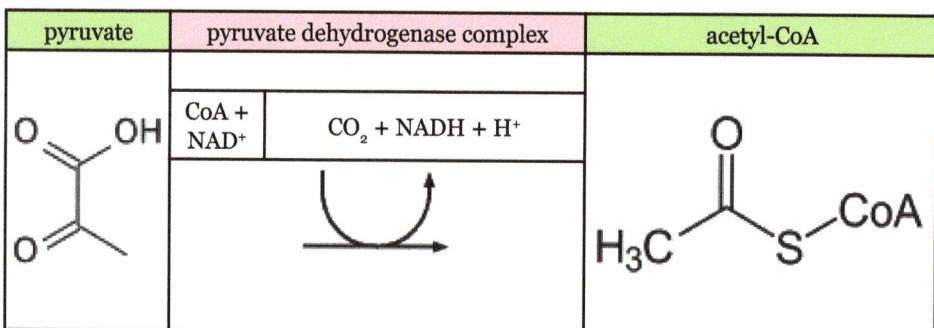

pyruvate	pyruvate dehydrogenase complex	acetyl-CoA
O, OH, O	CoA + NAD^+ → CO_2 + NADH + H^+	O, H_3C, S, CoA

Carboxylation to Oxaloacetate

Carboxylation by pyruvate carboxylase produces oxaloacetate.

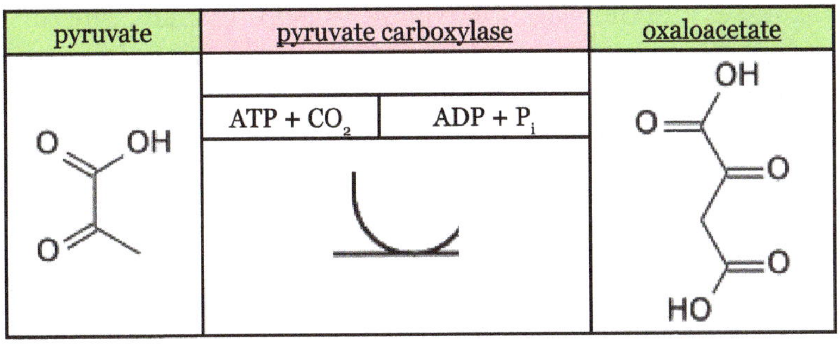

pyruvate	pyruvate carboxylase		oxaloacetate
	$ATP + CO_2$	$ADP + P_i$	

Transamination to Alanine

Transamination by alanine transaminase produces alanine.

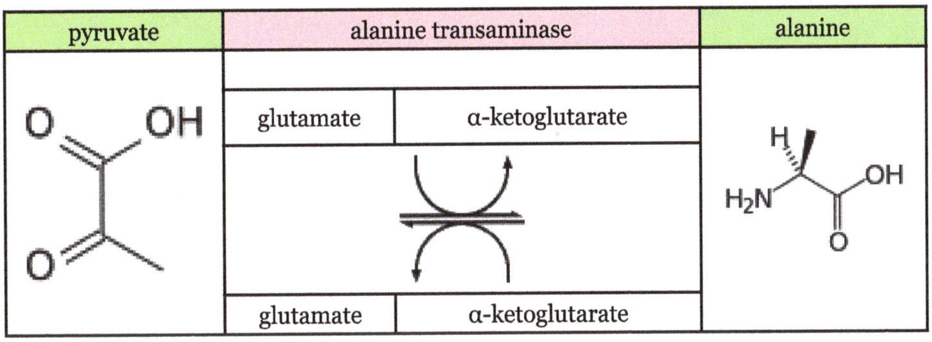

pyruvate	alanine transaminase		alanine
	glutamate	α-ketoglutarate	
	glutamate	α-ketoglutarate	

Reduction to Lactate

Reduction by lactate dehydrogenase produces lactate.

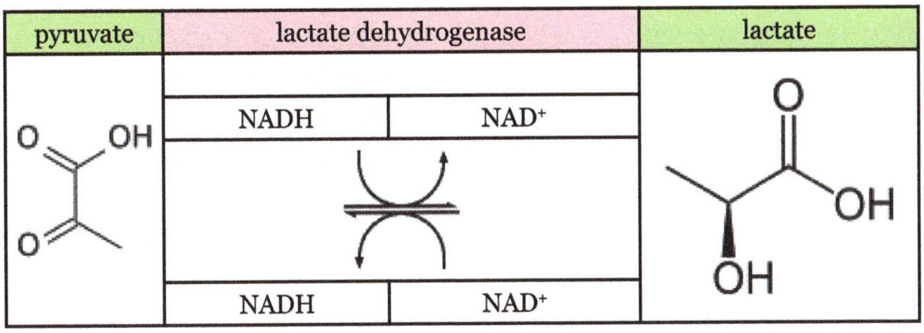

pyruvate	lactate dehydrogenase		lactate
	NADH	NAD^+	
	NADH	NAD^+	

Uses

The pyruvic acid derivative bromopyruvic acid is being studied for potential cancer treatment applications by researchers at Johns Hopkins University in ways that would support the Warburg hypothesis on the cause(s) of cancer.

Pyruvate is sold as a weight-loss supplement, though evidence supporting this use is lacking. A systematic review of six trials found a statistically significant difference in body weight with pyruvate compared to placebo. However, all of the trials had methodological weaknesses and the magnitude of the effect was small. The review also identified adverse events associated with pyruvate such as diarrhea, bloating, gas, and increase in low-density lipoprotein (LDL) cholesterol. The authors concluded that there was insufficient evidence to support the use of pyruvate for weight loss.

There is also *in vitro* evidence in hearts that pyruvate improves metabolism by NADH production stimulation and increases cardiac function.

Uvitic acid is obtained by condensing pyroracemic acid with baryta water.

Acetyl-CoA

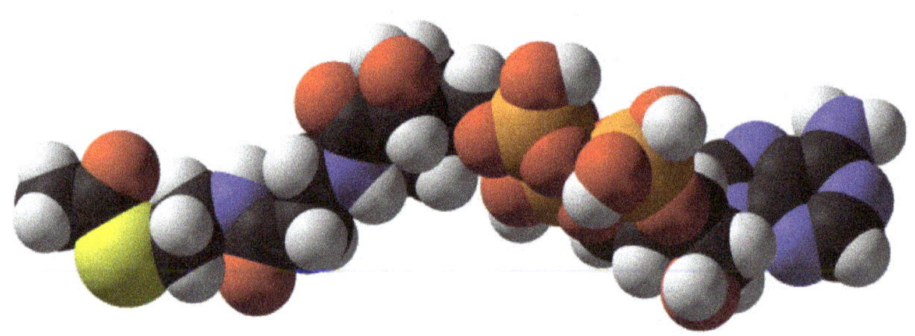

Acetyl CoA Space-filled Model

Acetyl coenzyme A or acetyl-CoA is an important molecule in metabolism, used in many biochemical reactions. Its main function is to deliver the acetyl group to the citric acid cycle (Krebs cycle) to be oxidized for energy production. Coenzyme A (CoASH or CoA) consists of a β-mercaptoethylamine group linked to the vitamin pantothenic acid through an amide linkage. The acetyl group (indicated in blue in the structural diagram on the right) of acetyl-CoA is linked to the sulfhydryl substituent of the β-mercaptoethylamine group. This thioester linkage is a "high energy" bond, which is particularly reactive. Hydrolysis of the thioester bond is exergonic (-31.5 kJ/mol).

CoA is acetylated during breakdown of carbohydrates through glycolysis, as well as by the beta-oxidation of fatty acids. It then enters the citric acid cycle, where the acetyl group is further oxidized to carbon dioxide and water, with the energy thus released captured in the form of 11 ATP and 1 GTP per acetyl group.

Konrad Bloch and Feodor Lynen were awarded the 1964 Nobel Prize in Physiology and Medicine for their discoveries linking acetyl-CoA and fatty acid metabolism. Fritz Lipmann won the Nobel Prize in 1953 for his discovery of the cofactor Coenzyme A.

Direct Synthesis

The acetylation of CoA is determined by the carbon sources.

Extramitochondrial:

- Under high glucose level, glycolysis takes place rapidly, thus increasing the amount of citrate produced from TCA cycle. This citrate is then exported to other cellular compartments outside mitochondria to be cleaved into acetyl-CoA and oxaloacetate by the enzyme ATP-citrate lyase (ACL). This reaction is coupled with the hydrolysis of ATP.
- Under low glucose level:
    - CoA is acetylated by acetyl-CoA synthetase (ACS), also coupled with ATP hydrolysis.
    - Ethanol also serves as a carbon source for acetylation of CoA utilizing the enzyme alcohol dehydrogenase.
    - Degradation of branched-chain ketogenic amino acids such as valine, leucine, and isoleucine. These amino acids are converted to alpha keto-acid by transamination and eventually to isovaleryl-CoA through oxidative decarboxylation by alpha-ketoacid dehydrogenase complex. Isovaleryl CoA undergoes dehydrogenation, carboxylation and hydration to form another CoA-derivative intermediate before it is cleaved into acetyl CoA and acetoacetate.

Intramitochondrial:

- Under high glucose level, acetyl CoA is produced through glycolysis. Pyruvate undergoes oxidative decarboxylation in which it loses its carboxyl group (as carbon dioxide) to form acetyl CoA, giving off 33.5kJ/mol.The oxidative conversion of pyruvate into acetyl-CoA is referred to as the pyruvate dehydrogenase reaction. It is catalyzed by the pyruvate dehydrogenase complex. Other conversions between pyruvate and acetyl-CoA are possible. For example, pyruvate formate lyase disproportionates pyruvate into acetyl-CoA and formic acid.
- Under low glucose level, the production of Acetyl CoA is linked to Beta oxidation of Fatty acid. Fatty acids are first converted to acyl-CoA. Acyl-CoA is then degraded in a 4-step cycle of dehydrogenation, hydration, oxidation and thiolysis catalyzed by four respective enzymes namely Acyl CoA Dehydrogenase, Enoyl CoA Hydratase, 3-hydroxyacyl CoA dehydrogenase, and Thiolase. The cycle produces a new Acyl-CoA with two less carbons and Acetyl CoA as byproduct .

pyruvate + CoA-SH + NAD^+ → acetyl CoA + NADH + CO_2

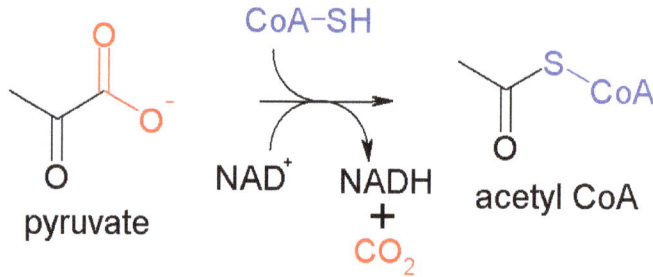

Pyruvate dehydrogenase complex reaction

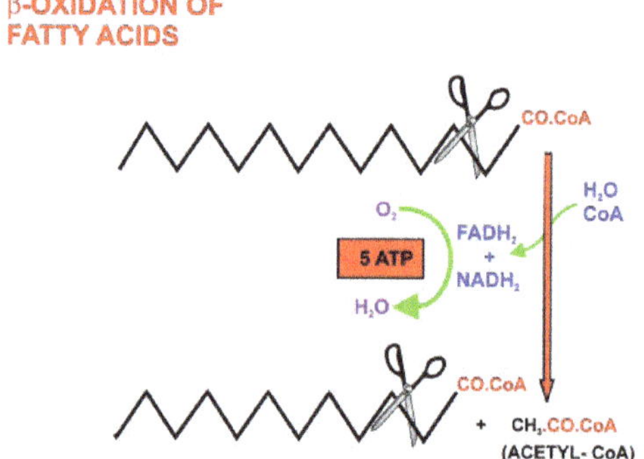

Functions

Intermediates In Various Pathways:

- Citric acid cycle:
    - Acetyl CoA reacts with oxaloacetate to form citrate, which is then oxidized to CO_2 in the cycle.
- Fatty acid metabolism
    - Acetyl-CoA is produced by the breakdown of both carbohydrates (by glycolysis) and fats (by beta-oxidation). It then enters the citric acid cycle in the mitochondrion by combining with oxaloacetate to form citrate.

Two acetyl-CoA molecules condense to form acetoacetyl-CoA, which gives rise to the formation of acetoacetate and beta-hydroxybutyrate. Acetoacetate, beta-hydroxybutyrate, and their spontaneous breakdown product, acetone, are frequently, but confusingly, known as ketone bodies (as they are not "bodies" at all, but water-soluble chemical substances). The ketone bodies are released by the liver into the blood. All cells with mitochondria can take ketone bodies up from the blood and reconvert them into acetyl-CoA, which can then be used as fuel in their citric acid cycles, as no other tissue can divert its oxaloacetate into the gluconeogenic pathway in the way that the liver does. Unlike free fatty acids, ketone bodies can cross the blood-brain barrier and are therefore available as fuel for the cells of the central nervous system, acting as a substitute for glucose, on which these cells normally survive. The occurrence of high levels of ketone bodies in the blood during starvation, a low carbohydrate diet, prolonged heavy exercise, and uncontrolled type-1 diabetes mellitus is known as ketosis, and in its extreme form in out-of-control type-1 diabetes mellitus, as ketoacidosis.

- On the other hand, when the insulin concentration in the blood is high, and that of glucagon is low (i.e. after meals), the acetyl-CoA produced by *glycolysis* condenses as normal with oxaloacetate to form citrate in the mitochondrion. However, instead of continuing through the citric acid cycle to be converted to carbon dioxide and water, the citrate is removed from the mitochondrion into the cytoplasm. There it is cleaved by ATP citrate lyase into acetyl-CoA and oxaloacetate. The oxaloacetate is returned to mitochondrion as malate (and then converted back into oxaloacetate to transfer more acetyl-CoA out of the mitochondrion). This cytosolic acetyl-CoA can then be used to synthesize fatty acids through carboxylation by acetyl CoA carboxylase into malonyl CoA, the first committed step in the synthesis of fatty acids. This conversion occurs primarily in the liver, adipose tissue and lactating mammary glands, where the fatty acids are combined with glycerol to form triglycerides, the major fuel reservoir of most animals. Fatty acids are also components of the phospholipids that make up the bulk of the lipid bilayers of all the cellular membranes.

- In plants, *de novo* fatty acid synthesis occurs in the plastids. Many seeds accumulate large reservoirs of seed oils to support germination and early growth of the seedling before it is a net photosynthetic organism.

- The cytosolic acetyl-CoA can also condense with acetoacetyl-CoA to form 3-hydroxy-3-methylglutaryl-CoA (HMG-CoA) which is the rate limiting step controlling the synthesis of cholesterol. Cholesterol can be used as is, as a structural component of cellular membranes, or it can be used to synthesize the steroid hormones, bile salts, and vitamin D.

- Acetyl-CoA can be carboxylated in the cytosol by acetyl-CoA carboxylase, giving rise to malonyl-CoA, a substrate required for synthesis of flavonoids and related polyketides, for elongation of fatty acids to produce waxes, cuticle, and seed oils in members of the Brassica family, and for malonation of proteins and other phytochemicals. In plants, these include sesquiterpenes, brassinosteroids (hormones), and membrane sterols.

- Steroid synthesis:

- Acetyl CoA participates in Mevalonate pathway by partaking in the synthesis of hydroxymethyl glutaryl-CoA

- Acetylcholine synthesis:
  - Acetyl-CoA is also an important component in the biogenic synthesis of the neurotransmitter acetylcholine. Choline, in combination with acetyl-CoA, is catalyzed by the enzyme choline acetyltransferase to produce acetylcholine and a Coenzyme A byproduct.
- Melatonin synthesis

Acetylation:

- Acetyl-CoA is also the source of the acetyl group incorporated onto certain lysine residues of histone and nonhistone proteins in the posttranslational modification acetylation. This acetylation is catalyzed by acetyltransferases. This acetylation affects cell growth, mitosis, and apoptosis

Allosteric regulator:

- Acetyl CoA serves as an allosteric regulator of pyruvate dehydrogenase kinase (PDH). It regulates through the ratio of acetyl-CoA versus CoA. Increased concentration of Acetyl CoA activates PDH.
- Acetyl CoA is also an allosteric activator of Pyruvate carboxylase

Citric Acid Cycle

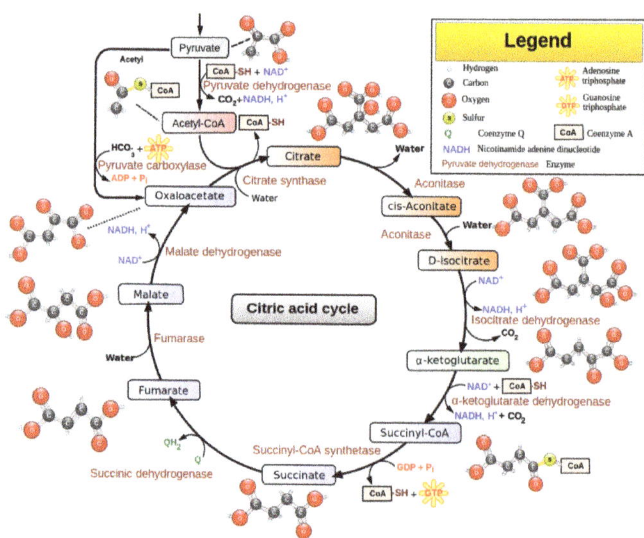

Overview of the citric acid cycle (click to enlarge)

The citric acid cycle – also known as the tricarboxylic acid (TCA) cycle or the Krebs cycle – is a series of chemical reactions used by all aerobic organisms to generate energy through the oxidation of acetyl-CoA derived from carbohydrates, fats and proteins into carbon dioxide and chemical energy in the form of adenosine triphosphate. In addition, the cycle provides precursors of certain amino acids as well as the reducing agent NADH that is used in numerous other biochemical reactions. Its central importance to many biochemical pathways suggests that it was one of the earliest established components of cellular metabolism and may have originated abiogenically.

The name of this metabolic pathway is derived from citric acid (a type of tricarboxylic acid) that is consumed and then regenerated by this sequence of reactions to complete the cycle. In addition, the cycle consumes acetate (in the form of acetyl-CoA) and water, reduces NAD^+ to NADH, and produces carbon dioxide as a waste byproduct. The NADH generated by the TCA cycle is fed into the oxidative phosphorylation (electron transport) pathway. The net result of these two closely linked pathways is the oxidation of nutrients to produce usable chemical energy in the form of ATP.

In eukaryotic cells, the citric acid cycle occurs in the matrix of the mitochondrion. In prokaryotic cells, such as bacteria which lack mitochondria, the TCA reaction sequence is performed in the cytosol with the proton gradient for ATP production being across the cell's surface (plasma membrane) rather than the inner membrane of the mitochondrion.

Discovery

Several of the components and reactions of the citric acid cycle were established in the 1930s by the research of the Nobel laureate Albert Szent-Györgyi, for which he received the Nobel Prize in 1937 for his discoveries pertaining to fumaric acid, a key component of the cycle. The citric acid cycle itself was finally identified in 1937 by Hans Adolf Krebs while at the University of Sheffield, for which he received the Nobel Prize for Physiology or Medicine in 1953.

Evolution

Components of the TCA cycle were derived from anaerobic bacteria, and the TCA cycle itself may have evolved more than once. Theoretically there are several alternatives to the TCA cycle; however, the TCA cycle appears to be the most efficient. If several TCA alternatives had evolved independently, they all appear to have converged to the TCA cycle.

Overview

The citric acid cycle is a key metabolic pathway that unifies carbohydrate, fat, and protein metabolism. The reactions of the cycle are carried out by 8 enzymes that completely oxidize acetate, in the form of acetyl-CoA, into two molecules each of carbon dioxide and water. Through catabolism of sugars, fats, and proteins, a two-carbon organic product acetate in the form of acetyl-CoA is produced which enters the citric acid cycle. The reactions of the cycle also convert three equivalents of nicotinamide adenine dinucleotide (NAD^+) into three equivalents of reduced NAD^+ (NADH), one equivalent of flavin adenine dinucleotide (FAD) into one equivalent of $FADH_2$, and one equivalent each of guanosine diphosphate (GDP) and inorganic phosphate (P_i) into one equivalent of guanosine triphosphate (GTP). The NADH and $FADH_2$ generated by the citric acid cycle are in turn used by the oxidative phosphorylation pathway to generate energy-rich adenosine triphosphate (ATP).

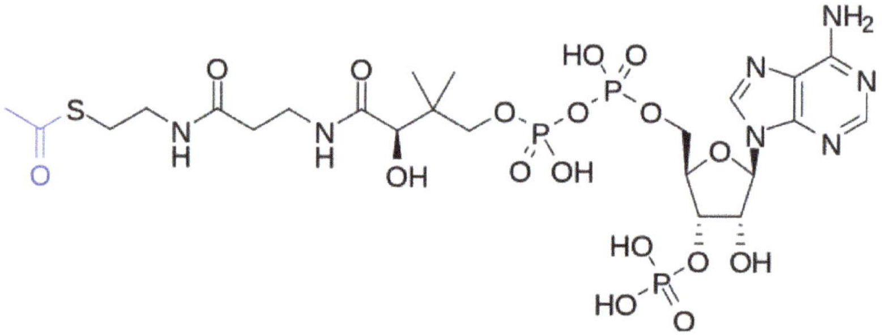

Structural diagram of acetyl-CoA. The portion in blue, on the left, is the acetyl group; the portion in black is coenzyme A.

One of the primary sources of acetyl-CoA is from the breakdown of sugars by glycolysis which yield pyruvate that in turn is decarboxylated by the enzyme pyruvate dehydrogenase generating acetyl-CoA according to the following reaction scheme:

- $CH_3C(=O)C(=O)O^-$ (pyruvate) + HSCoA + NAD^+ → $CH_3C(=O)SCoA$ (acetyl-CoA) + NADH + CO_2

The product of this reaction, acetyl-CoA, is the starting point for the citric acid cycle. Acetyl-CoA may also be obtained from the oxidation of fatty acids. Below is a schematic outline of the cycle:

- The citric acid cycle begins with the transfer of a two-carbon acetyl group from acetyl-CoA to the four-carbon acceptor compound (oxaloacetate) to form a six-carbon compound (citrate).
- The citrate then goes through a series of chemical transformations, losing two carboxyl groups as CO_2. The carbons lost as CO_2 originate from what was oxaloacetate, not directly from acetyl-CoA. The carbons donated by acetyl-CoA become part of the oxaloacetate carbon backbone after the first turn of the citric acid cycle. Loss of the acetyl-CoA-donated carbons as CO_2 requires several turns of the citric acid cycle. However, because of the role of the citric acid cycle in anabolism, they might not be lost, since many TCA cycle intermediates are also used as precursors for the biosynthesis of other molecules.
- Most of the energy made available by the oxidative steps of the cycle is transferred as energy-rich electrons to NAD^+, forming NADH. For each acetyl group that enters the citric acid cycle, three molecules of NADH are produced.

- Electrons are also transferred to the electron acceptor Q, forming QH_2.
- At the end of each cycle, the four-carbon oxaloacetate has been regenerated, and the cycle continues.

Steps

Two carbon atoms are oxidized to CO_2, the energy from these reactions being transferred to other metabolic processes by GTP (or ATP), and as electrons in NADH and QH_2. The NADH generated in the TCA cycle may later donate its electrons [oxidative phosphorylation] to drive ATP synthesis; $FADH_2$ is covalently attached to succinate dehydrogenase, an enzyme functioning both in the TCA cycle and the mitochondrial electron transport chain in oxidative phosphorylation. $FADH_2$, therefore, facilitates transfer of electrons to coenzyme Q, which is the final electron acceptor of the reaction catalyzed by the Succinate:ubiquinone oxidoreductase complex, also acting as an intermediate in the electron transport chain.

The citric acid cycle is continuously supplied with new carbon in the form of acetyl-CoA, entering at step 0 below.

	Substrates	**Products**	**Enzyme**	**Reaction type**	**Comment**
0 / 10	Oxaloacetate +Acetyl CoA +H_2O	Citrate +CoA-SH	Citrate synthase	Aldol condensation	irreversible,extends the 4C oxaloacetate to a 6C molecule
1	Citrate	*cis*-Aconitate +H_2O	Aconitase	Dehydration	reversible isomerisation
2	*cis*-Aconitate +H_2O	Isocitrate		Hydration	
3	Isocitrate +NAD^+	Oxalosuccinate +NADH + H^+	Isocitrate dehydrogenase	Oxidation	generates NADH (equivalent of 2.5 ATP)
4	Oxalosuccinate	α-Ketoglutarate +CO_2		Decarboxylation	rate-limiting, irreversible stage, generates a 5C molecule
5	α-Ketoglutarate +NAD^+ +CoA-SH	Succinyl-CoA +NADH + H^+ +CO_2	α-Ketoglutarate dehydrogenase	Oxidative-decarboxylation	irreversible stage, generates NADH (equivalent of 2.5 ATP),regenerates the 4C chain (CoA excluded)
6	Succinyl-CoA +GDP + P_i	Succinate +CoA-SH +GTP	Succinyl-CoA synthetase	substrate-level phosphorylation	or ADP→ATP instead of GDP→GTP,generates 1 ATP or equivalent Condensation reaction of GDP + P_i and hydrolysis of Succinyl-CoA involve the H_2O needed for balanced equation.

7	Succinate +ubiquinone (Q)	Fumarate +ubiquinol (QH_2)	Succinate dehydrogenase	Oxidation	uses FAD as a prosthetic group ($FAD \rightarrow FADH_2$ in the first step of the reaction) in the enzyme,-generates the equivalent of 1.5 ATP
8	Fumarate +H_2O	*L*-Malate	Fumarase	Hydration	Hydration of C-C double bond
9	*L*-Malate +NAD^+	Oxaloacetate +NADH + H^+	Malate dehydrogenase	Oxidation	reversible (in fact, equilibrium favors malate), generates NADH (equivalent of 2.5 ATP)
10 / 0	Oxaloacetate +Acetyl CoA +H_2O	Citrate +CoA-SH	Citrate synthase	Aldol condensation	This is the same as step 0 and restarts the cycle. The reaction is irreversible and extends the 4C oxaloacetate to a 6C molecule

Mitochondria in animals, including humans, possess two succinyl-CoA synthetases: one that produces GTP from GDP, and another that produces ATP from ADP. Plants have the type that produces ATP (ADP-forming succinyl-CoA synthetase). Several of the enzymes in the cycle may be loosely associated in a multienzyme protein complex within the mitochondrial matrix.

The GTP that is formed by GDP-forming succinyl-CoA synthetase may be utilized by nucleoside-diphosphate kinase to form ATP (the catalyzed reaction is GTP + ADP → GDP + ATP).

Products

Products of the first turn of the cycle are: *one GTP (or ATP), three NADH, one* QH_2, *two* CO_2.

Because two acetyl-CoA molecules are produced from each glucose molecule, two cycles are required per glucose molecule. Therefore, at the end of two cycles, the products are: two GTP, six NADH, two QH_2, and four CO_2

Description	Reactants	Products
The sum of all reactions in the citric acid cycle is:	Acetyl-CoA + 3 NAD^+ + Q + GDP + P_i + 2 H_2O	→ CoA-SH + 3 NADH + 3 H^+ + QH_2 + GTP + 2 CO_2
Combining the reactions occurring during the pyruvate oxidation with those occurring during the citric acid cycle, the following overall pyruvate oxidation reaction is obtained:	Pyruvate ion + 4 NAD^+ + Q + GDP + P_i + 2 H_2O	→ 4 NADH + 4 H^+ + QH_2 + GTP + 3 CO_2

Combining the above reaction with the ones occurring in the course of glycolysis, the following overall glucose oxidation reaction (excluding reactions in the respiratory chain) is obtained:	Glucose + 10 NAD^+ + 2 Q + 2 ADP + 2 GDP + 4 P_i + 2 H_2O	→ 10 NADH + 10 H^+ + 2 QH_2 + 2 ATP + 2 GTP + 6 CO_2

The above reactions are balanced if P_i represents the $H_2PO_4^-$ ion, ADP and GDP the ADP^{2-} and GDP^{2-} ions, respectively, and ATP and GTP the ATP^{3-} and GTP^{3-} ions, respectively.

The total number of ATP obtained after complete oxidation of one glucose in glycolysis, citric acid cycle, and oxidative phosphorylation is estimated to be between 30 and 38.

Efficiency

The theoretical maximum yield of ATP through oxidation of one molecule of glucose in glycolysis, citric acid cycle, and oxidative phosphorylation is 38 (assuming 3 molar equivalents of ATP per equivalent NADH and 2 ATP per $FADH_2$). In eukaryotes, two equivalents of NADH are generated in glycolysis, which takes place in the cytoplasm. Transport of these two equivalents into the mitochondria consumes two equivalents of ATP, thus reducing the net production of ATP to 36. Furthermore, inefficiencies in oxidative phosphorylation due to leakage of protons across the mitochondrial membrane and slippage of the ATP synthase/proton pump commonly reduces the ATP yield from NADH and $FADH_2$ to less than the theoretical maximum yield. The observed yields are, therefore, closer to ~2.5 ATP per NADH and ~1.5 ATP per $FADH_2$, further reducing the total net production of ATP to approximately 30. An assessment of the total ATP yield with newly revised proton-to-ATP ratios provides an estimate of 29.85 ATP per glucose molecule.

Variation

While the TCA cycle is in general highly conserved, there is significant variability in the enzymes found in different taxa (note that the diagrams on this page are specific to the mammalian pathway variant).

Some differences exist between eukaryotes and prokaryotes. The conversion of D-*threo*-isocitrate to 2-oxoglutarate is catalyzed in eukaryotes by the NAD^+-dependent EC 1.1.1.41, while prokaryotes employ the $NADP^+$-dependent EC 1.1.1.42. Similarly, the conversion of (*S*)-malate to oxaloacetate is catalyzed in eukaryotes by the NAD^+-dependent EC 1.1.1.37, while most prokaryotes utilize a quinone-dependent enzyme, EC 1.1.5.4.

A step with significant variability is the conversion of succinyl-CoA to succinate. Most organisms utilize EC 6.2.1.5, succinate–CoA ligase (ADP-forming) (despite its name, the enzyme operates in the pathway in the direction of ATP formation). In mammals a GTP-forming enzyme, succinate–CoA ligase (GDP-forming) (EC 6.2.1.4) also operates.

The level of utilization of each isoform is tissue dependent. In some acetate-producing bacteria, such as *Acetobacter aceti*, an entirely different enzyme catalyzes this conversion – EC 2.8.3.18, succinyl-CoA:acetate CoA-transferase. This specialized enzyme links the TCA cycle with acetate metabolism in these organisms. Some bacteria, such as *Helicobacter pylori*, employ yet another enzyme for this conversion – succinyl-CoA:acetoacetate CoA-transferase (EC 2.8.3.5).

Some variability also exists at the previous step – the conversion of 2-oxoglutarate to succinyl-CoA. While most organisms utilize the ubiquitous NAD^+-dependent 2-oxoglutarate dehydrogenase, some bacteria utilize a ferredoxin-dependent 2-oxoglutarate synthase (EC 1.2.7.3). Other organisms, including obligately autotrophic and methanotrophic bacteria and archaea, bypass succinyl-CoA entirely, and convert 2-oxoglutarate to succinate via succinate semialdehyde, using EC 4.1.1.71, 2-oxoglutarate decarboxylase, and EC 1.2.1.79, succinate-semialdehyde dehydrogenase.

Regulation

The regulation of the TCA cycle is largely determined by product inhibition and substrate availability. If the cycle were permitted to run unchecked, large amounts of metabolic energy could be wasted in overproduction of reduced coenzyme such as NADH and ATP. The major eventual substrate of the cycle is ADP which gets converted to ATP. A reduced amount of ADP causes accumulation of precursor NADH which in turn can inhibit a number of enzymes. NADH, a product of all dehydrogenases in the TCA cycle with the exception of succinate dehydrogenase, inhibits pyruvate dehydrogenase, isocitrate dehydrogenase, α-ketoglutarate dehydrogenase, and also citrate synthase. Acetyl-coA inhibits pyruvate dehydrogenase, while succinyl-CoA inhibits alpha-ketoglutarate dehydrogenase and citrate synthase. When tested in vitro with TCA enzymes, ATP inhibits citrate synthase and α-ketoglutarate dehydrogenase; however, ATP levels do not change more than 10% in vivo between rest and vigorous exercise. There is no known allosteric mechanism that can account for large changes in reaction rate from an allosteric effector whose concentration changes less than 10%.

Calcium is used as a regulator. Mitochondrial matrix calcium levels can reach the tens of micromolar levels during cellular activation. It activates pyruvate dehydrogenase phosphatase which in turn activates the pyruvate dehydrogenase complex. Calcium also activates isocitrate dehydrogenase and α-ketoglutarate dehydrogenase. This increases the reaction rate of many of the steps in the cycle, and therefore increases flux throughout the pathway.

Citrate is used for feedback inhibition, as it inhibits phosphofructokinase, an enzyme involved in glycolysis that catalyses formation of fructose 1,6-bisphosphate, a precursor of pyruvate. This prevents a constant high rate of flux when there is an accumulation of citrate and a decrease in substrate for the enzyme.

Recent work has demonstrated an important link between intermediates of the citric acid cycle and the regulation of hypoxia-inducible factors (HIF). HIF plays a role in the regulation of oxygen homeostasis, and is a transcription factor that targets angiogenesis, vascular remodeling, glucose utilization, iron transport and apoptosis. HIF is synthesized consititutively, and hydroxylation of at least one of two critical proline residues mediates their interaction with the von Hippel Lindau E3 ubiquitin ligase complex, which targets them for rapid degradation. This reaction is catalysed by prolyl 4-hydroxylases. Fumarate and succinate have been identified as potent inhibitors of prolyl hydroxylases, thus leading to the stabilisation of HIF.

Major Metabolic Pathways Converging on the TCA Cycle

Several catabolic pathways converge on the TCA cycle. Most of these reactions add intermediates to the TCA cycle, and are therefore known as anaplerotic reactions, from the Greek meaning to "fill up". These increase the amount of acetyl CoA that the cycle is able to carry, increasing the mitochondrion's capability to carry out respiration if this is otherwise a limiting factor. Processes that remove intermediates from the cycle are termed "cataplerotic" reactions.

In this section and in the next, the citric acid cycle intermediates are indicated in *italics* to distinguish them from other substrates and end-products.

Pyruvate molecules produced by glycolysis are actively transported across the inner mitochondrial membrane, and into the matrix. Here they can be oxidized and combined with coenzyme A to form CO_2, *acetyl-CoA*, and NADH, as in the normal cycle.

However, it is also possible for pyruvate to be carboxylated by pyruvate carboxylase to form *oxaloacetate*. This latter reaction "fills up" the amount of *oxaloacetate* in the citric acid cycle, and is therefore an anaplerotic reaction, increasing the cycle's capacity to metabolize *acetyl-CoA* when the tissue's energy needs (e.g. in muscle) are suddenly increased by activity.

In the citric acid cycle all the intermediates (e.g. *citrate, iso-citrate, alpha-ketoglutarate, succinate, fumarate, malate* and *oxaloacetate*) are regenerated during each turn of the cycle. Adding more of any of these intermediates to the mitochondrion therefore means that that additional amount is retained within the cycle, increasing all the other intermediates as one is converted into the other. Hence the addition of any one of them to the cycle has an anaplerotic effect, and its removal has a cataplerotic effect. These anaplerotic and cataplerotic reactions will, during the course of the cycle, increase or decrease the amount of *oxaloacetate* available to combine with *acetyl-CoA* to form *citric acid*. This in turn increases or decreases the rate of ATP production by the mitochondrion, and thus the availability of ATP to the cell.

Acetyl-CoA, on the other hand, derived from pyruvate oxidation, or from the beta-oxidation of fatty acids, is the only fuel to enter the citric acid cycle. With each turn of the cycle one molecule of *acetyl-CoA* is consumed for every molecule of *oxaloacetate* present in the mitochondrial matrix, and is never regenerated. It is the oxidation of the acetate portion of *acetyl-CoA* that produces CO_2 and water, with the energy thus released captured in the form of ATP.

In the liver, the carboxylation of cytosolic pyruvate into intra-mitochondrial *oxaloacetate* is an early step in the gluconeogenic pathway which converts lactate and de-aminated alanine into glucose, under the influence of high levels of glucagon and/or epinephrine in the blood. Here the addition of *oxaloacetate* to the mitochondrion does not have a net anaplerotic effect, as another citric acid cycle intermediate (*malate*) is immediately removed from the mitochondrion to be converted into cytosolic oxaloacetate, which is ultimately converted into glucose, in a process that is almost the reverse of glycolysis.

In protein catabolism, proteins are broken down by proteases into their constituent amino acids. Their carbon skeletons (i.e. the de-aminated amino acids) may either enter the citric acid cycle as intermediates (e.g. *alpha-ketoglutarate* derived from glutamate or glutamine), having an anaplerotic effect on the cycle, or, in the case of leucine, isoleucine, lysine, phenylalanine, tryptophan, and tyrosine, they are converted into *acetyl-CoA* which can be burned to CO_2 and water, or used to form ketone bodies, which too can only be burned in tissues other than the liver where they are formed, or excreted via the urine or breath. These latter amino acids are therefore termed "ketogenic" amino acids, whereas those that enter the citric acid cycle as intermediates can only be cataplerotically removed by entering the gluconeogenic pathway via *malate* which is transported out of the mitochondrion to be converted into cytosolic oxaloacetate and ultimately into glucose. These are the so-called "glucogenic" amino acids. De-aminated alanine, cysteine, glycine, serine, and threonine are converted to pyruvate and can consequently either enter the citric acid cycle as *oxaloacetate* (an anaplerotic reaction) or as *acetyl-CoA* to be disposed of as CO_2 and water.

In fat catabolism, triglycerides are hydrolyzed to break them into fatty acids and glycerol. In the liver the glycerol can be converted into glucose via dihydroxyacetone phosphate and glyceraldehyde-3-phosphate by way of gluconeogenesis. In many tissues, especially heart and skeletal muscle tissue, fatty acids are broken down through a process known as beta oxidation, which results in the production of mitochondrial *acetyl-CoA*, which can be used in the citric acid cycle. Beta oxidation of fatty acids with an odd number of methylene bridges produces propionyl-CoA, which is then converted into *succinyl-CoA* and fed into the citric acid cycle as an anaplerotic intermediate.

The total energy gained from the complete breakdown of one (six-carbon) molecule of glucose by glycolysis, the formation of 2 *acetyl-CoA* molecules, their catabolism in the citric acid cycle, and oxidative phosphorylation equals about 30 ATP molecules, in eukaryotes. The number of ATP molecules derived from the beta oxidation of a 6 carbon segment of a fatty acid chain, and the subsequent oxidation of the resulting 3 molecules of *acetyl-CoA* is 40.

Citric Acid Cycle Intermediates Serve as Substrates for Biosynthetic Processes

In this subheading, as in the previous one, the TCA intermediates are identified by *italics*.

Several of the citric acid cycle intermediates are used for the synthesis of important compounds, which will have significant cataplerotic effects on the cycle. *Acetyl-CoA* cannot be transported out of the mitochondrion. To obtain cytosolic acetyl-CoA, *citrate* is removed from the citric acid cycle and carried across the inner mitochondrial membrane into the cytosol. There it is cleaved by ATP citrate lyase into acetyl-CoA and oxaloacetate. The oxaloacetate is returned to mitochondrion as *malate* (and then converted back into *oxaloacetate* to transfer more *acetyl-CoA* out of the mitochondrion). The cytosolic acetyl-CoA is used for fatty acid synthesis and the production of cholesterol. Cholesterol can, in turn, be used to synthesize the steroid hormones, bile salts, and vitamin D.

The carbon skeletons of many non-essential amino acids are made from citric acid cycle intermediates. To turn them into amino acids the alpha keto-acids formed from the citric acid cycle intermediates have to acquire their amino groups from glutamate in a transamination reaction, in which pyridoxal phosphate is a cofactor. In this reaction the glutamate is converted into *alpha-ketoglutarate*, which is a citric acid cycle intermediate. The intermediates that can provide the carbon skeletons for amino acid synthesis are *oxaloacetate* which forms aspartate and asparagine; and *alpha-ketoglutarate* which forms glutamine, proline, and arginine.

Of these amino acids, aspartate and glutamine are used, together with carbon and nitrogen atoms from other sources, to form the purines that are used as the bases in DNA and RNA, as well as in ATP, AMP, GTP, NAD, FAD and CoA.

The pyrimidines are partly assembled from aspartate (derived from *oxaloacetate*). The pyrimidines, thymine, cytosine and uracil, form the complementary bases to the purine bases in DNA and RNA, and are also components of CTP, UMP, UDP and UTP.

The majority of the carbon atoms in the porphyrins come from the citric acid cycle intermediate, *succinyl-CoA*. These molecules are an important component of the hemoproteins, such as hemoglobin, myoglobin and various cytochromes.

During gluconeogenesis mitochondrial *oxaloacetate* is reduced to *malate* which is then transported out of the mitochondrion, to be oxidized back to oxaloacetate in the cytosol. Cytosolic oxaloacetate is then decarboxylated to phosphoenolpyruvate by phosphoenolpyruvate carboxykinase, which is the rate limiting step in the conversion of nearly all the gluconeogenic precursors (such as the glucogenic amino acids and lactate) into glucose by the liver and kidney.

Because the citric acid cycle is involved in both catabolic and anabolic processes, it is known as an amphibole pathway.

References

- Lehninger, Albert (2008). Principles of Biochemistry. New York, NY: W.H. Freeman and Company. p. 539. ISBN 0-7167-7108-X.
- Kurtzman CP, Fell JW. (2005). Biodiversity and Ecophysiology of Yeasts (in: The Yeast Handbook, Gábor P, de la Rosa CL, eds.). Berlin: Springer. pp. 11–30. ISBN 978-3-540-26100-1.
- Snodgrass ME. (2004). Encyclopedia of Kitchen History. New York, New York: Fitzroy Dearborn. p. 1066. ISBN 978-1-57958-380-4.
- Kaufmann K, Schoneck A. (2002). Making Sauerkraut and Pickled Vegetables at Home: Creative Recipes for Lactic Fermented Food to Improve Your Health. Book Publishing Company. ISBN 978-1-55312-037-7.
- Bernstein H, Bernstein C. (2013). "Evolutionary Origin and Adaptive Function of Meiosis". In Bernstein C; Bernstein H. Meiosis. ISBN 978-953-51-1197-9. Retrieved 29 May 2016.
- Gibson M. (2010). The Sommelier Prep Course: An Introduction to the Wines, Beers, and Spirits of the World. John Wiley and Sons. p. 361. ISBN 978-0-470-28318-9.
- Amendola J, Rees N. (2002). Understanding Baking: The Art and Science of Baking. John Wiley and Sons. p. 36. ISBN 978-0-471-40546-7.
- Smith A, Kraig B. (2013). The Oxford Encyclopedia of Food and Drink in America. Oxford University Press. p. 440. ISBN 978-0-19-973496-2.
- Thaler M, Safferstein D. (2014). A Curious Harvest: The Practical Art of Cooking Everything. Quarry Books. p. 129. ISBN 978-1-59253-928-4.
- Duyff RL. (2012). American Dietetic Association Complete Food and Nutrition Guide, Revised and Updated (4th ed.). Houghton Mifflin Harcourt. pp. 256–257. ISBN 978-0-544-66456-2.
- Downes FP, Ito K. (2001). Compendium of Methods for the Microbiological Examination of Foods. Washington, DC: American Public Health Association. p. 211. ISBN 978-0-87553-175-5.
- Patterson MJ (1996). Baron S; et al., eds. Streptococcus. In: Baron's Medical Microbiology (4th ed.). Univ of Texas Medical Branch. (via NCBI Bookshelf) ISBN 0-9631172-1-1.
- Chatterjea (2004-01-01). Textbook of Biochemistry for Dental/Nursing/Pharmacy Students. Jaypee Brothers Publishers. ISBN 9788180612046.
- Berg, Jeremy M.; Tymoczko, John L.; Stryer, Lubert; Berg, Jeremy M.; Tymoczko, John L.; Stryer, Lubert (2002-01-01). Biochemistry (5th ed.). W H Freeman. ISBN 0716730510.
- Voet, Donald; Judith G. Voet; Charlotte W. Pratt (2006). Fundamentals of Biochemistry, 2nd Edition. John Wiley and Sons, Inc. pp. 547, 556. ISBN 0-471-21495-7.

5

Glycolysis: An Overview

Glycolysis is the oxygen dependent biochemical pathway that metabolizes glucose to produce pyruvate, energy, water and two positively charged hydrogen ions. This section explores the metabolic pathway of glycolysis and also delves into allosteric regulation and the role of the enzyme phosphofructokinase 1 in the process of glycolysis.

Glycolysis

Glycolysis (from *glycose*, an older term for glucose + *-lysis* degradation) is the metabolic pathway that converts glucose $C_6H_{12}O_6$, into pyruvate, $CH_3COCOO^- + H^+$. The free energy released in this process is used to form the high-energy molecules ATP (adenosine triphosphate) and NADH (reduced nicotinamide adenine dinucleotide).

Glycolysis is a determined sequence of ten enzyme-catalyzed reactions. The intermediates provide entry points to glycolysis. For example, most monosaccharides, such as fructose and galactose, can be converted to one of these intermediates. The intermediates may also be directly useful. For example, the intermediate dihydroxyacetone phosphate (DHAP) is a source of the glycerol that combines with fatty acids to form fat.

Glycolysis is an oxygen independent metabolic pathway, meaning that it does not use molecular oxygen (i.e. atmospheric oxygen) for any of its reactions. However the products of glycolysis (pyruvate and NADH + H^+) are sometimes metabolized using atmospheric oxygen. When molecular oxygen is used for the metabolism of the products of glycolysis the process is usually referred to as aerobic, whereas if no oxygen is used the process is said to be anaerobic. Thus, glycolysis occurs, with variations, in nearly all organisms, both aerobic and anaerobic. The wide occurrence of glycolysis indicates that it is one of the most ancient metabolic pathways. Indeed, the reactions that constitute glycolysis and its parallel pathway, the pentose phosphate pathway, occur metal-catalyzed under the oxygen-free conditions of the Archean oceans, also in the absence of enzymes. Glycolysis could thus have originated from chemical constraints of the prebiotic world.

Glycolysis occurs in most organisms in the cytosol of the cell. The most common type of glycolysis is the *Embden–Meyerhof–Parnas (EMP pathway)*, which was discovered by Gustav Embden, Otto Meyerhof, and Jakub Karol Parnas. Glycolysis also refers to other pathways, such as the *Entner–Doudoroff pathway* and various heterofermentative and homofermentative pathways. However, the discussion here will be limited to the Embden–Meyerhof–Parnas pathway.

The entire glycolysis pathway can be separated into two phases:

1. The Preparatory Phase – in which ATP is consumed and is hence also known as the investment phase
2. The Pay Off Phase – in which ATP is produced.

Overview

The overall reaction of glycolysis is:

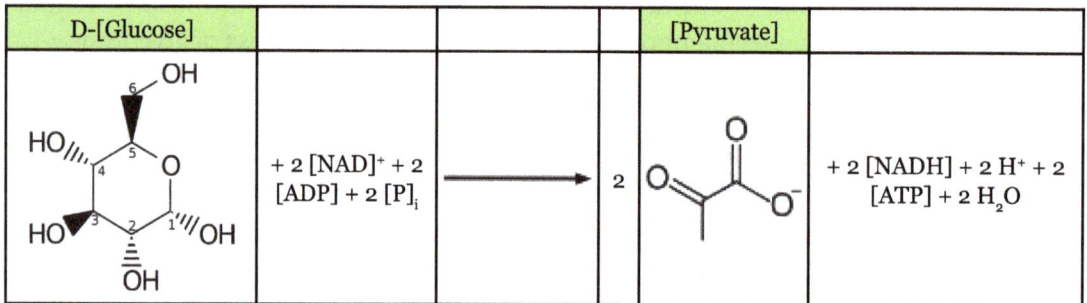

D-[Glucose]				[Pyruvate]	
HO, HO, OH, OH, O, OH (structure)	+ 2 $[NAD]^+$ + 2 [ADP] + 2 $[P]_i$	→	2	O, O, O^- (structure)	+ 2 [NADH] + 2 H^+ + 2 [ATP] + 2 H_2O

The use of symbols in this equation makes it appear unbalanced with respect to oxygen atoms, hydrogen atoms, and charges. Atom balance is maintained by the two phosphate (P_i) groups:

- Each exists in the form of a hydrogen phosphate anion (HPO_4^{2-}), dissociating to contribute 2 H^+ overall
- Each liberates an oxygen atom when it binds to an ADP (adenosine diphosphate) molecule, contributing 2 O overall

Charges are balanced by the difference between ADP and ATP. In the cellular environment, all three hydroxyl groups of ADP dissociate into $-O^-$ and H^+, giving ADP^{3-}, and this ion tends to exist in an ionic bond with Mg^{2+}, giving $ADPMg^-$. ATP behaves identically except that it has four hydroxyl groups, giving $ATPMg^{2-}$. When these differences along with the true charges on the two phosphate groups are considered together, the net charges of −4 on each side are balanced.

For simple fermentations, the metabolism of one molecule of glucose to two molecules of pyruvate has a net yield of two molecules of ATP. Most cells will then carry out further reactions to 'repay' the used NAD^+ and produce a final product of ethanol or lactic acid. Many bacteria use inorganic compounds as hydrogen acceptors to regenerate the NAD^+.

Cells performing aerobic respiration synthesize much more ATP, but not as part of glycolysis. These further aerobic reactions use pyruvate and NADH + H^+ from glycolysis. Eukaryotic aerobic respiration produces approximately 34 additional molecules of ATP for each glucose molecule, however most of these are produced by a vastly different mechanism to the substrate-level phosphorylation in glycolysis.

The lower-energy production, per glucose, of anaerobic respiration relative to aerobic respiration, results in greater flux through the pathway under hypoxic (low-oxygen) conditions, unless alternative sources of anaerobically oxidizable substrates, such as fatty acids, are found.

History

The pathway of glycolysis as it is known today took almost 100 years to fully discover. The combined results of many smaller experiments were required in order to understand the pathway as a whole.

The first steps in understanding glycolysis began in the nineteenth century with the wine industry. For economic reasons, the French wine industry sought to investigate why wine sometime turned distasteful, instead of fermenting into alcohol. French scientist Louis Pasteur researched this issue during the 1850s, and the results of his experiments began the long road to elucidating the pathway of glycolysis. His experiments showed that fermentation occurs by the action of living microorganisms; and that yeast's glucose consumption decreased under aerobic conditions of fermentation, in comparison to anaerobic conditions (the Pasteur Effect).

Eduard Buchner. Discovered cell-free fermentation.

While Pasteur's experiments were groundbreaking, insight into the component steps of glycolysis were provided by the non-cellular fermentation experiments of Eduard Buchner during the 1890s. Buchner demonstrated that the conversion of glucose to ethanol was possible using a non-living extract of yeast (due to the action of enzymes in the extract). This experiment not only revolutionized biochemistry, but also allowed later scientists to analyze this pathway in a more controlled lab setting. In a series of experiments (1905-1911), scientists Arthur Harden and William Young discovered more pieces of glycolysis. . They discovered the regulatory effects of ATP on glucose consumption during alcohol fermentation. They also shed light on the role of one compound as a glycolysis intermediate: fructose 1,6-bisphosphate.

The elucidation of Fructose 1,6-diphosphate was accomplished by measuring CO2 levels when yeast juice was incubated with glucose. CO2 production increased rapidly then slowed down. Harden and Young noted that this process would restart if an inorganic phosphate (Pi) was added to the mixture. Harden and Young deduced that this process produced organic phosphate esters, and further experiments allowed them to extract fructose diphosphate (F-1,6-DP).

Arthur Harden and William Young along with Nick Sheppard determined, in a second experiment, that a heat-sensitive high-molecular-weight subcellular fraction (the enzymes) and a heat-insensitive low-molecular-weight cytoplasm fraction (ADP, ATP and NAD^+ and other cofactors) are required together for fermentation to proceed. This experiment begun by observing that dialyzed (purified) yeast juice could not ferment or even create a sugar phosphate. This mixture was rescued with the addition of undialyzed yeast extract that had been boiled. Boiling the yeast extract renders all proteins inactive (as it denatures them). The ability of boiled extract plus dialyzed juice to complete fermentation suggests that the cofactors were non-protein in character.

Otto Meyerhof. One of the main scientists involved in completing the puzzle of glycolysis

In the 1920s Otto Meyerhof was able to link together some of the many individual pieces of glycolysis discovered by Buchner, Harden, and Young. Meyerhof and his team was able to extract different glycolytic enzymes from muscle tissue, and combine them to artificially create the pathway from glycogen to lactic acid.

In one paper, Meyerhof and scientist Renate Junowicz-Kockolaty investigated the reaction that splits fructose 1,6-diphosohate into the two triose phosphates. Previous work proposed that the split occurred via 1,3-diphosphoglyceraldehye plus an oxidizing enzyme and cozymase. Meyerhoff and Junowicz found that the equilibrium constant for the isomerase and aldoses reaction were not affected by inorganic phosphates or any other cozymase or oxidizing enzymes. They further removed diphosphoglyceraldehyde as a possible intermediate in glycolysis.

With all of these pieces available by the 1930s, Gustav Embden proposed a detailed, step-by-step outline of that pathway we now know as glycolysis. The biggest difficulties in determining the intricacies of the pathway were due to the very short lifetime and low steady-state concentrations of the intermediates of the fast glycolytic reactions. By the 1940s, Meyerhof, Embden and many other biochemists had finally completed the puzzle of glycolysis. The understanding of the isolated pathway has been expanded in the subsequent decades, to include further details of its regulation and integration with other metabolic pathways.

Sequence of Reactions

Glycolysis metabolic pathway

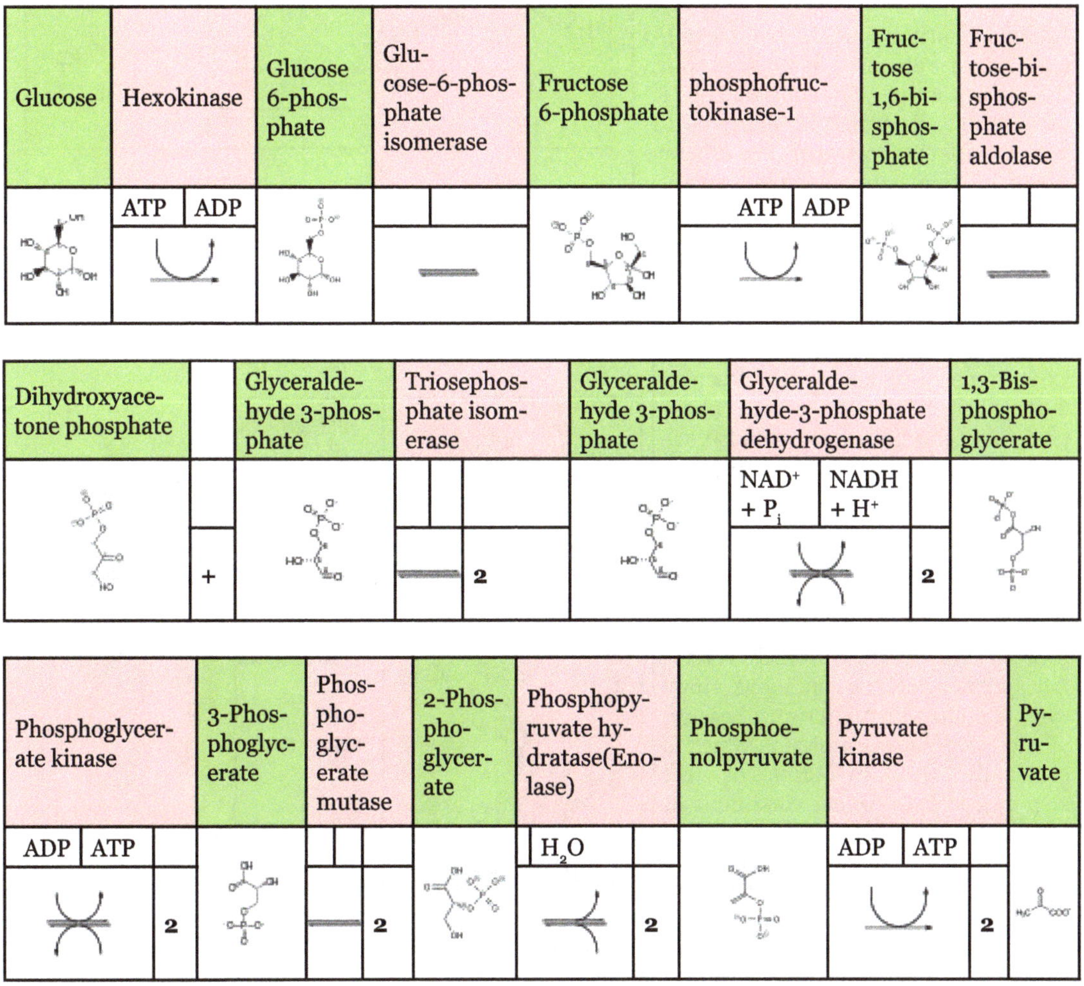

Glucose	Hexokinase	Glucose 6-phosphate	Glucose-6-phosphate isomerase	Fructose 6-phosphate	phosphofructokinase-1	Fructose 1,6-bisphosphate	Fructose-bisphosphate aldolase
	ATP ADP				ATP ADP		

Dihydroxyacetone phosphate		Glyceraldehyde 3-phosphate	Triosephosphate isomerase	Glyceraldehyde 3-phosphate	Glyceraldehyde-3-phosphate dehydrogenase	1,3-Bisphosphoglycerate
	+		2		NAD^+ + P_i NADH + H^+ 2	

Phosphoglycerate kinase	3-Phosphoglycerate	Phosphoglycerate mutase	2-Phosphoglycerate	Phosphopyruvate hydratase(Enolase)	Phosphoenolpyruvate	Pyruvate kinase	Pyruvate
ADP ATP 2		2		H_2O 2		ADP ATP 2	

Preparatory phase

The first five steps are regarded as the preparatory (or investment) phase, since they consume energy to convert the glucose into two three-carbon sugar phosphates (G3P).

The first step in glycolysis is phosphorylation of glucose by a family of enzymes called hexokinases to form glucose 6-phosphate (G6P). This reaction consumes ATP, but it acts to keep the glucose concentration low, promoting continuous transport of glucose into the cell through the plasma membrane transporters. In addition, it blocks the glucose from leaking out – the cell lacks transporters for G6P, and free diffusion out of the cell is prevented due to the charged nature of G6P. Glucose may alternatively be formed from the phosphorolysis or hydrolysis of intracellular starch or glycogen.

In animals, an isozyme of hexokinase called glucokinase is also used in the liver, which has a much lower affinity for glucose (K_m in the vicinity of normal glycemia), and differs in regulatory properties. The different substrate affinity and alternate regulation of this enzyme are a reflection of the role of the liver in maintaining blood sugar levels.

Cofactors: Mg

D-Glucose (**Glc**)	Hexokinase (**HK**) *a transferase*	α-D-Glucose-6-phosphate (**G6P**)
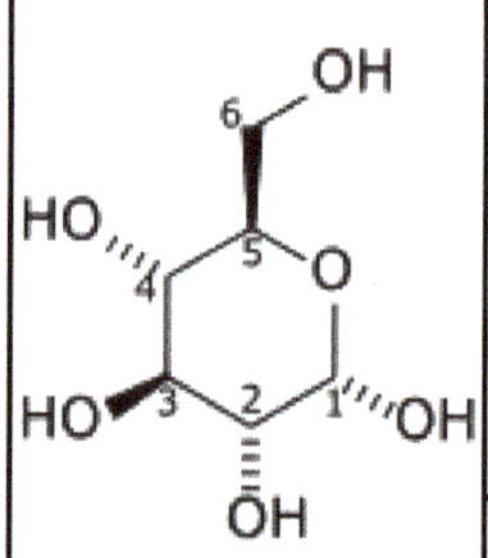	ATP → H^+ + ADP 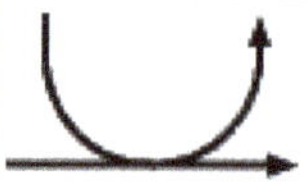	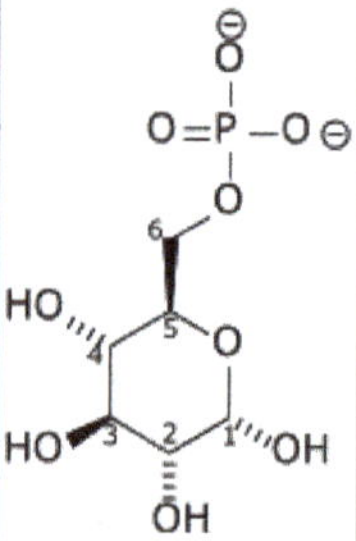

G6P is then rearranged into fructose 6-phosphate (F6P) by glucose phosphate isomerase. Fructose can also enter the glycolytic pathway by phosphorylation at this point.

The change in structure is an isomerization, in which the G6P has been converted to F6P. The reaction requires an enzyme, phosphohexose isomerase, to proceed. This reaction is freely reversible under normal cell conditions. However, it is often driven forward because of a low concentration of F6P, which is constantly consumed during the next step of glycolysis. Under conditions of high F6P concentration, this reaction readily runs in reverse. This phenomenon can be explained through Le Chatelier's Principle. Isomerization to a keto sugar is necessary for carbanion stabilization in the fourth reaction step (below).

α-D-Glucose 6-phosphate (**G6P**)	Phosphoglucose isomerase (**PGI**) *an isomerase*	β-D-Fructose 6-phosphate (**F6P**)
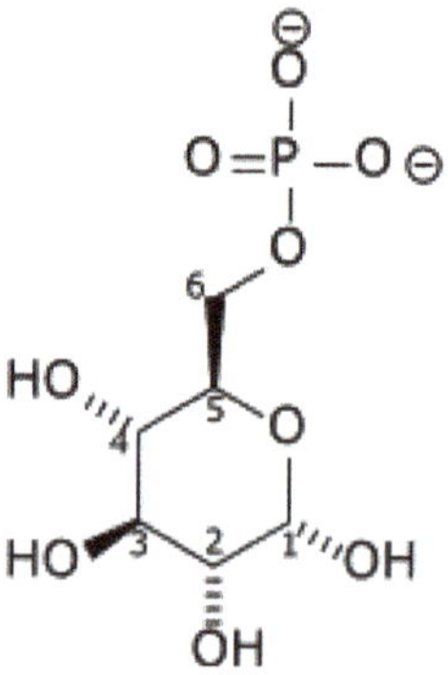	 ⇌	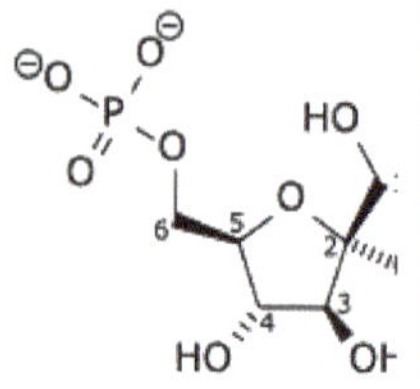

The energy expenditure of another ATP in this step is justified in 2 ways: The glycolytic process (up to this step) is now irreversible, and the energy supplied destabilizes the molecule. Because the reaction catalyzed by Phosphofructokinase 1 (PFK-1) is coupled to the hydrolysis of ATP (an energetically favorable step) it is, in essence, irreversible, and a different pathway must be used to do the reverse conversion during gluconeogenesis. This makes the reaction a key regulatory point. This is also the rate-limiting step.

Furthermore, the second phosphorylation event is necessary to allow the formation of two charged groups (rather than only one) in the subsequent step of glycolysis, ensuring the prevention of free diffusion of substrates out of the cell.

The same reaction can also be catalyzed by pyrophosphate-dependent phosphofructokinase (PFP or PPi-PFK), which is found in most plants, some bacteria, archea, and protists, but not in animals. This enzyme uses pyrophosphate (PPi) as a phosphate donor instead of ATP. It is a reversible reaction, increasing the flexibility of glycolytic metabolism. A rarer ADP-dependent PFK enzyme variant has been identified in archaean species.

Cofactors: Mg

β-D-Fructose 6-phosphate **(F6P)**	phosphofructokinase **(PFK-1)** *a transferase*	β-D-Fructose 1,6-bisphosphate **(F1,6BP)**
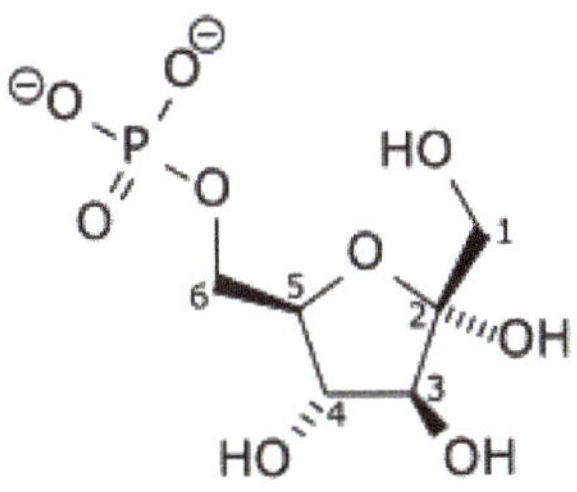	ATP → H^+ + ADP 	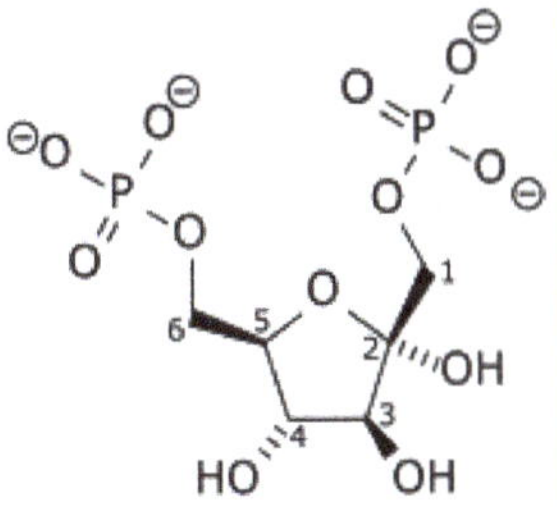

Destabilizing the molecule in the previous reaction allows the hexose ring to be split by aldolase into two triose sugars,dihydroxyacetone phosphate, a ketose, and glyceraldehyde 3-phosphate, an aldose. There are two classes of aldolases: class I aldolases, present in animals and plants, and class II aldolases, present in fungi and bacteria; the two classes use different mechanisms in cleaving the ketose ring.

Electrons delocalized in the carbon-carbon bond cleavage associate with the alcohol group. The resulting carbanion is stabilized by the structure of the carbanion itself via resonance charge distribution and by the presence of a charged ion prosthetic group.

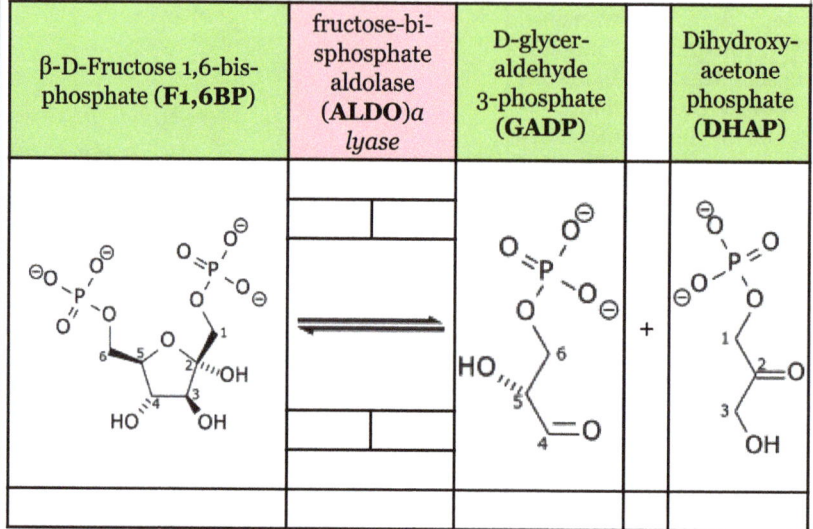

β-D-Fructose 1,6-bisphosphate (**F1,6BP**)	fructose-bisphosphate aldolase (**ALDO**)*a lyase*	D-glyceraldehyde 3-phosphate (**GADP**)		Dihydroxyacetone phosphate (**DHAP**)
			+	

Triosephosphate isomerase rapidly interconverts dihydroxyacetone phosphate with glyceraldehyde 3-phosphate (GADP) that proceeds further into glycolysis. This is advantageous, as it directs dihydroxyacetone phosphate down the same pathway as glyceraldehyde 3-phosphate, simplifying regulation.

Dihydroxyacetone phosphate (**DHAP**)	triosephosphate isomerase (**TPI**) *an isomerase*	D-glyceraldehyde 3-phosphate (**GADP**)

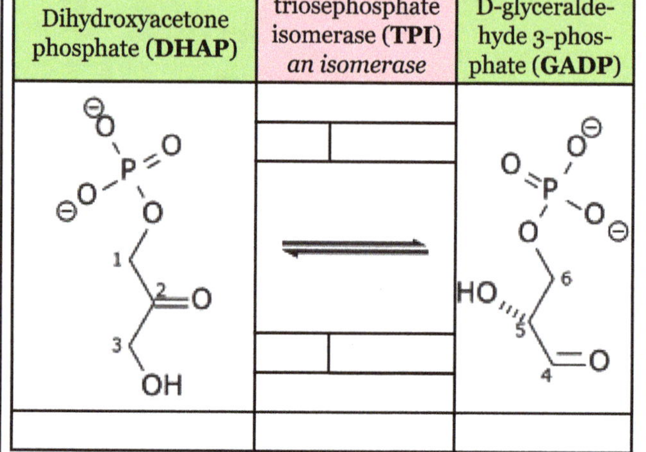

Pay-off Phase

The second half of glycolysis is known as the pay-off phase, characterised by a net gain of the energy-rich molecules ATP and NADH. Since glucose leads to two triose sugars in the preparatory phase, each reaction in the pay-off phase occurs twice per glucose molecule. This yields 2 NADH molecules and 4 ATP molecules, leading to a net gain of 2 NADH molecules and 2 ATP molecules from the glycolytic pathway per glucose.

The aldehyde groups of the triose sugars are oxidised, and inorganic phosphate is added to them, forming 1,3-bisphosphoglycerate.

The hydrogen is used to reduce two molecules of NAD^+, a hydrogen carrier, to give NADH + H^+ for each triose.

Hydrogen atom balance and charge balance are both maintained because the phosphate (P_i) group actually exists in the form of a hydrogen phosphate anion (HPO_4^{2-}), which dissociates to contribute the extra H^+ ion and gives a net charge of -3 on both sides.

glyceraldehyde 3-phosphate **(GADP)**	glyceraldehyde phosphate dehydrogenase **(GAPDH)** *oxidoreductase*	D-1,3-bisphosphoglycerate **(1,3BPG)**
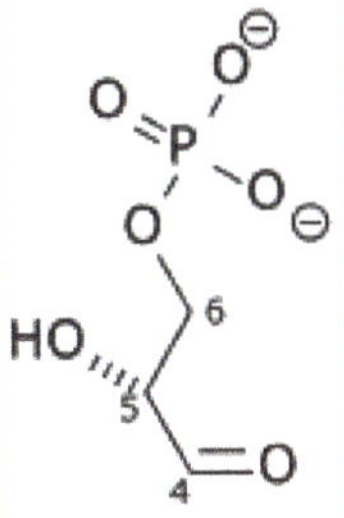	NAD^+ + P_i → NADH + H^+ 	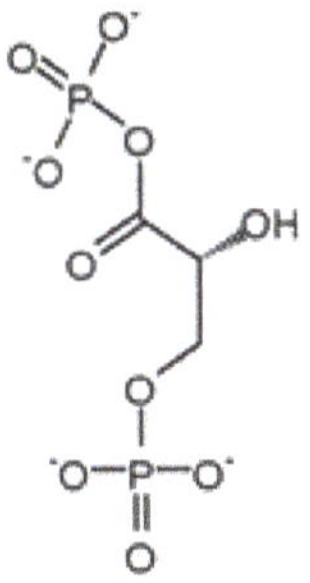

This step is the enzymatic transfer of a phosphate group from 1,3-bisphosphoglycerate to ADP by phosphoglycerate kinase, forming ATP and 3-phosphoglycerate. At this step, glycolysis has reached the break-even point: 2 molecules of ATP were consumed, and 2 new molecules have now been synthesized. This step, one of the two substrate-level phosphorylation steps, requires ADP; thus, when the cell has plenty of ATP (and little ADP), this reaction does not occur. Because ATP decays relatively quickly when it is not metabolized, this is an important regulatory point in the glycolytic pathway.

ADP actually exists as $ADPMg^-$, and ATP as $ATPMg^{2-}$, btalancing the charges at -5 both sides.

Cofactors: Mg

1,3-bisphosphoglycerate **(1,3BPG)**	phosphoglycerate kinase **(PGK)***a transferase*	3-phosphoglycerate **(3PG)**
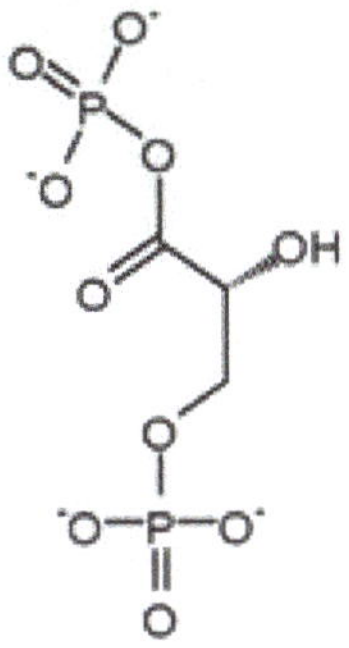	ADP → ATP 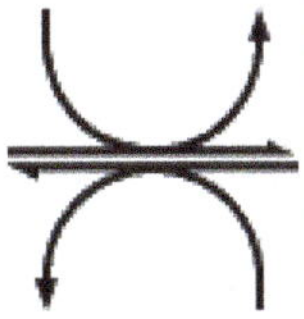	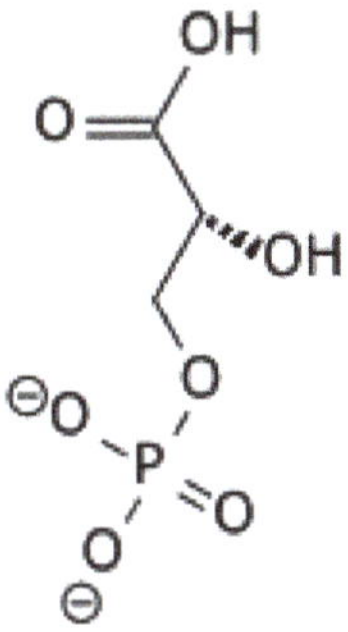
	phosphoglycerate kinase **(PGK)**	

	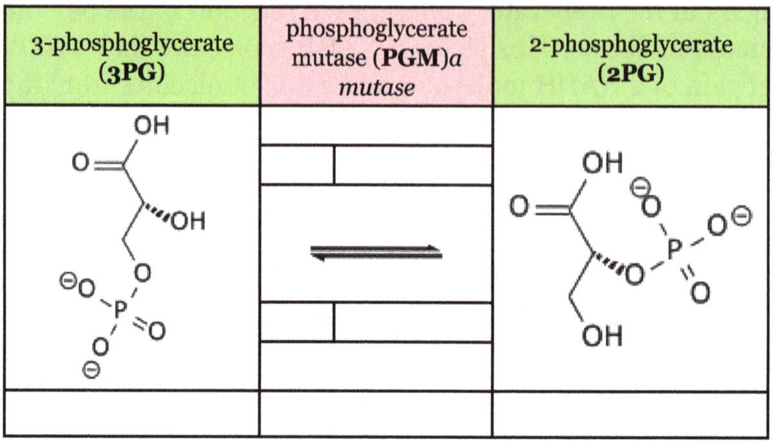 3-phosphoglycerate (**3PG**)	phosphoglycerate mutase (**PGM**)*a mutase*	2-phosphoglycerate (**2PG**)
Phosphoglycerate mutase isomerises 3-phosphoglycerate into 2-phosphoglycerate.	OH, O, OH, ⊖O, P, O, O⊖	⇌	OH, ⊖O, O⊖, O, P, O, OH

	2-phosphoglycerate (**2PG**)	enolase (**ENO**) *a lyase*	phosphoenolpyruvate (**PEP**)
Enolase next forms phosphoenolpyruvate from 2-phosphoglycerate. *Cofactors:* 2 Mg^{2+}: one "conformational" ion to coordinate with the carboxylate group of the substrate, and one "catalytic" ion that participates in the dehydration.	OH, O, ⊖O, O⊖, P, O, OH	H_2O ⇌	O, OH, O, ⊖O–P=O, O⊖
		enolase (**ENO**)	

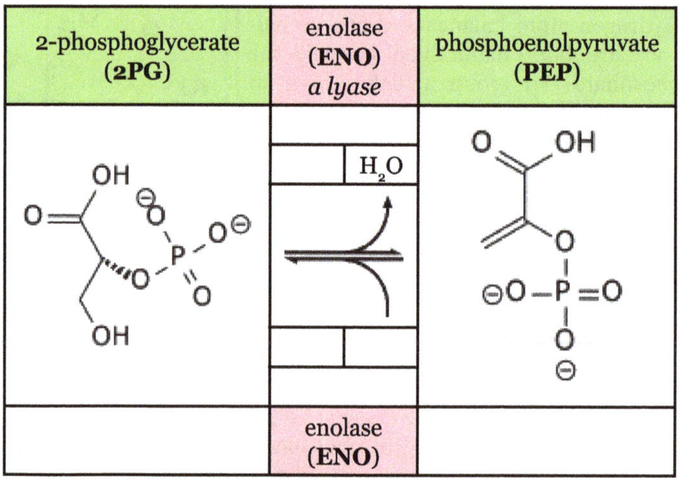

	phosphoenolpyruvate (**PEP**)	pyruvate kinase (**PK**) *a transferase*	pyruvate (**Pyr**)
A final substrate-level phosphorylation now forms a molecule of pyruvate and a molecule of ATP by means of the enzyme pyruvate kinase. This serves as an additional regulatory step, similar to the phosphoglycerate kinase step. *Cofactors:* Mg	O, OH, O, ⊖O–P=O, O⊖	ADP + H^+ → ATP	O, OH, O

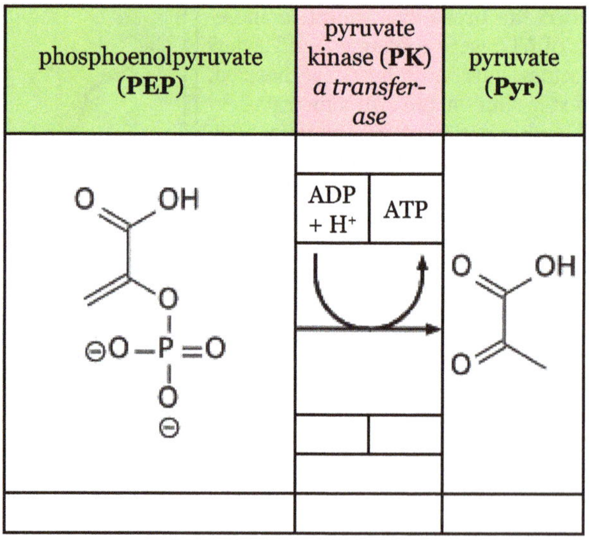

Regulation

Glycolysis is regulated by slowing down or speeding up certain steps in the pathway by inhibiting or activating the enzymes that are involved. The steps that are regulated may be determined by calculating the change in free energy, ΔG, for each step.

When ΔG is negative, a reaction proceeds spontaneously in the forward direction only and is considered irreversible. When ΔG is positive, the reaction is non-spontaneous and will not proceed in the forward direction unless coupled with an energetically favorable reaction. When ΔG is zero, the reaction is at equilibrium, can proceed in either directions and is considered reversible.

If a step is at equilibrium (ΔG is zero), the enzyme catalyzing the reaction will balance the products and reactants and cannot confer directionality to the pathway. These steps (and associated enzymes) are considered unregulated. If a step is not at equilibrium, but spontaneous (ΔG is negative), the enzyme catalyzing the reaction is not balancing the products and reactants and will proceed in the forward direction unless the enzyme function is altered; these steps are considered to be regulated. A common mechanism of regulating enzymes is allosteric control.

Free Energy Changes

Concentrations of metabolites in erythrocytes	
Compound	**Concentration / m*M***
Glucose	5.0
Glucose-6-phosphate	0.083
Fructose-6-phosphate	0.014
Fructose-1,6-bisphosphate	0.031
Dihydroxyacetone phosphate	0.14
Glyceraldehyde-3-phosphate	0.019
1,3-Bisphosphoglycerate	0.001
2,3-Bisphosphoglycerate	4.0
3-Phosphoglycerate	0.12
2-Phosphoglycerate	0.03
Phosphoenolpyruvate	0.023
Pyruvate	0.051
ATP	1.85
ADP	0.14
P_i	1.0

The change in free energy, ΔG, for each step in the glycolysis pathway can be calculated using $\Delta G = \Delta G^{\circ\prime} + RT\ln Q$, where Q is the reaction quotient. This requires knowing the concentrations of the metabolites. All of these values are available for erythrocytes, with the exception of the concentrations of NAD^+ and NADH. The ratio of NAD^+ to NADH in the cytoplasm is approximately 1000, which makes the oxidation of glyceraldehyde-3-phosphate (step 6) more favourable.

Using the measured concentrations of each step, and the standard free energy changes, the actual free energy change can be calculated. (Neglecting this is very common - the delta G of ATP hydrolysis in cells is not the standard free energy change of ATP hydrolysis quoted in textbooks).

Change in free energy for each step of glycolysis			
Step	**Reaction**	**$\Delta G^{\circ\prime}$ / (kJ/mol)**	**ΔG / (kJ/mol)**
1	Glucose + ATP^{4-} → Glucose-6-phosphate^{2-} + ADP^{3-} + H^+	-16.7	-34
2	Glucose-6-phosphate^{2-} → Fructose-6-phosphate^{2-}	1.67	-2.9
3	Fructose-6-phosphate^{2-} + ATP^{4-} → Fructose-1,6-bisphosphate^{4-} + ADP^{3-} + H^+	-14.2	-19
4	Fructose-1,6-bisphosphate^{4-} → Dihydroxyacetone phosphate^{2-} + Glyceraldehyde-3-phosphate^{2-}	23.9	-0.23
5	Dihydroxyacetone phosphate^{2-} → Glyceraldehyde-3-phosphate^{2-}	7.56	2.4
6	Glyceraldehyde-3-phosphate^{2-} + P_i^{2-} + NAD^+ → 1,3-Bisphosphoglycerate^{4-} + NADH + H^+	6.30	-1.29
7	1,3-Bisphosphoglycerate^{4-} + ADP^{3-} → 3-Phosphoglycerate^{3-} + ATP^{4-}	-18.9	0.09
8	3-Phosphoglycerate^{3-} → 2-Phosphoglycerate^{3-}	4.4	0.83
9	2-Phosphoglycerate^{3-} → Phosphoenolpyruvate^{3-} + H_2O	1.8	1.1
10	Phosphoenolpyruvate^{3-} + ADP^{3-} + H^+ → Pyruvate$^-$ + ATP^{4-}	-31.7	-23.0

From measuring the physiological concentrations of metabolites in an erythrocyte it seems that about seven of the steps in glycolysis are in equilibrium for that cell type. Three of the steps — the ones with large negative free energy changes — are not in equilibrium and are referred to as *irreversible*; such steps are often subject to regulation.

Step 5 in the figure is shown behind the other steps, because that step is a side-reaction that can decrease or increase the concentration of the intermediate glyceraldehyde-3-phosphate. That compound is converted to dihydroxyacetone phosphate by the enzyme triose phosphate isomerase, which is a catalytically perfect enzyme; its rate is so fast that the reaction can be assumed to be in equilibrium. The fact that ΔG is not zero indicates that the actual concentrations in the erythrocyte are not accurately known.

Biochemical Logic

The existence of more than one point of regulation indicates that intermediates between those points enter and leave the glycolysis pathway by other processes. For example, in the first regulated step, hexokinase converts glucose into glucose-6-phosphate. Instead of continuing through the glycolysis pathway, this intermediate can be converted into glucose storage molecules, such as glycogen or starch. The reverse reaction, breaking down, e.g., glycogen, produces mainly glucose-6-phosphate; very little free glucose is formed in the reaction. The glucose-6-phosphate so produced can enter glycolysis *after* the first control point.

In the second regulated step (the third step of glycolysis), phosphofructokinase converts fructose-6-phosphate into fructose-1,6-bisphosphate, which then is converted into glyceraldehyde-3-phosphate and dihydroxyacetone phosphate. The dihydroxyacetone phosphate can be removed from glycolysis by conversion into glycerol-3-phosphate, which can be used to form triglycerides. On the converse, triglycerides can be broken down into fatty acids and glycerol; the latter, in turn, can be converted into dihydroxyacetone phosphate, which can enter glycolysis *after* the second control point.

Regulation of the Rate Limiting Enzymes

The four regulatory enzymes are hexokinase, glucokinase, phosphofructokinase, and pyruvate kinase. The flux through the glycolytic pathway is adjusted in response to conditions both inside and outside the cell. The internal factors that regulate glycolysis do so primarily to provide ATP in adequate quantities for the cell's needs. The external factors act primarily on the liver, fat tissue, and muscles, which can remove large quantities of glucose from the blood after meals (thus preventing hyperglycemia by storing the excess glucose as fat or glycogen, depending on the tissue type). The liver is also capable of releasing glucose into the blood between meals, during fasting, and exercise thus preventing hypoglycemia by means of glycogenolysis and gluconeogenesis. These latter reactions coincide with the halting of glycolysis in the liver.

In animals, regulation of blood glucose levels by the pancreas in conjunction with the liver is a vital part of homeostasis. The beta cells in the pancreatic islets are sensitive to the blood glucose concentration. A rise in the blood glucose concentration causes them to release insulin into the blood, which has an effect particularly on the liver, but also on fat and muscle cells, causing these tissues to remove glucose from the blood. When the blood sugar falls the pancreatic beta cells cease insulin production, but, instead, stimulate the neighboring pancreatic alpha cells to release glucagon into the blood. This, in turn, causes the liver to release glucose into the blood by breaking down stored glycogen, and by means of gluconeogenesis. If the fall in the blood glucose level is particularly rapid or severe, other glucose sensors cause the release of epinephrine from the adrenal glands into the blood. This has the same action as glucagon on glucose metabolism, but its effect is more pronounced. In the liver glucagon and epinephrine cause the phosphorylation of the key, rate limiting enzymes of glycolysis, fatty acid synthesis, cholesterol synthesis, gluconeogenesis, and glycogenolysis. Insulin has the opposite effect on these enzymes. The phosphorylation and dephosphorylation of these enzymes (ultimately in response to the glucose level in the blood) is the dominant manner by which these pathways are controlled in the liver, fat, and muscle cells. Thus the phosphorylation of phosphofructokinase inhibits glycolysis, whereas its dephosphorylation through the action of insulin stimulates glycolysis.

In addition hexokinase and glucokinase act independently of the hormonal effects as controls at the entry points of glucose into the cells of different tissues. Hexokinase responds to the glucose-6-phosphate (G6P) level in the cell, or, in the case of glucokinase, to the blood sugar level in the blood to impart entirely intracellular controls of the glycolytic pathway in different tissues.

When glucose has been converted into G6P by hexokinase or glucokinase, it can either be converted to glucose-1-phosphate (G1P) for conversion to glycogen, or it is alternatively converted by glycolysis to pyruvate, which enters the mitochondrion where it is converted into acetyl-CoA and then into citrate. Excess citrate is exported from the mitochondrion back into to the cytosol, where ATP citrate lyase regenerates acetyl-CoA and oxaloacetate (OAA). The acetyl-CoA is then used for fatty acid synthesis and cholesterol synthesis, two important ways of utilizing excess glucose when its concentration is high in blood. The rate limiting enzymes catalyzing these reactions perform these functions when they have been dephosphorylated through the action of insulin on the liver cells. Between meals, during fasting, exercise or hypoglycemia, glucagon and epinephrine are released into the blood. This causes liver glycogen to be converted back to G6P, and then converted to glucose by the liver-specific enzyme glucose 6-phosphatase and released into the blood. Glucagon and epinephrine also stimulate gluconeogenesis, which coverts non-carbohydrate substrates into G6P, which joins the G6P derived from glycogen, or substitutes for it when the liver glycogen store have been depleted. This is critical for brain function, since the brain utilizes glucose as an energy source under most conditions. The simultaneously phosphorylation of, particularly, phosphofructokinase, but also, to a certain extent pyruvate kinase, prevents glycolysis occurring at the same time as gluconeogenesis and glycogenolysis.

Hexokinase and Glucokinase

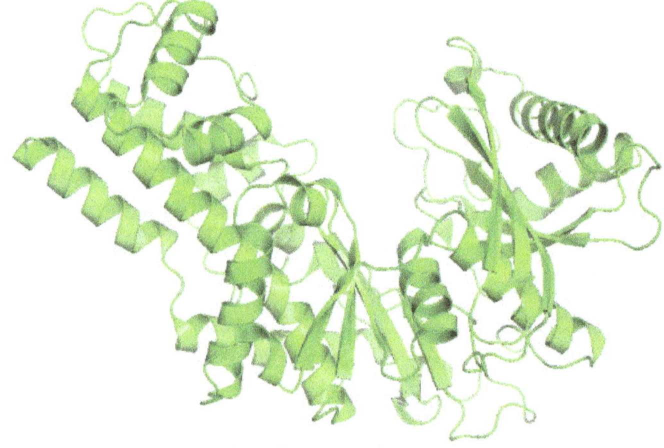

Yeast hexokinase B (PDB: 1IG8)

All cells contain the enzyme hexokinase, which catalyzes the conversion of glucose that has entered the cell into glucose-6-phosphate (G6P). Since the cell wall is impervious to G6P, hexokinase essentially acts to transport glucose into the cells from which it can then no longer escape. Hexokinase is inhibited by high levels of G6P in the cell. Thus the rate of entry of glucose into cells partially depends on how fast G6P can be disposed of by glycolysis, and by glycogen synthesis (in the cells which store glycogen, namely liver and muscles).

Glucokinase, unlike hexokinase, is not inhibited by G6P. It occurs in liver cells, and will only phosphorylate the glucose entering the cell to form glucose-6-phosphate (G6P), when the sugar in the blood is abundant. This being the first step in the glycolytic pathway in the liver, it therefore imparts an additional layer of control of the glycolytic pathway in this organ.

Phosphofructokinase

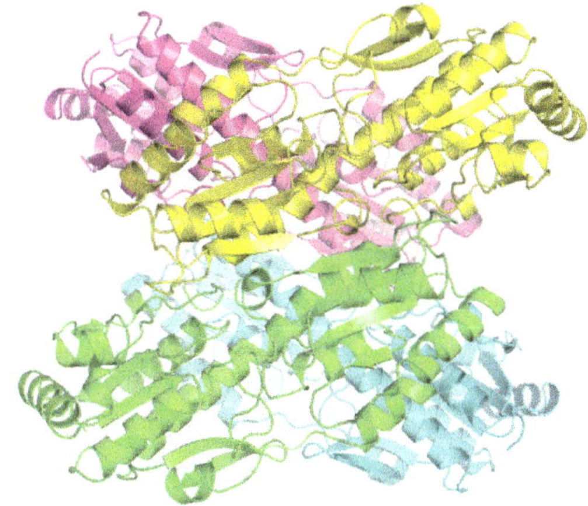

Bacillus stearothermophilus phosphofructokinase (PDB: 6PFK)

Phosphofructokinase is an important control point in the glycolytic pathway, since it is one of the irreversible steps and has key allosteric effectors, AMP and fructose 2,6-bisphosphate (F2,6BP).

Fructose 2,6-bisphosphate (F2,6BP) is a very potent activator of phosphofructokinase (PFK-1) that is synthesized when F6P is phosphorylated by a second phosphofructokinase (PFK2). In liver, when blood sugar is low and glucagon elevates cAMP, PFK2 is phosphorylated by protein kinase A. The phosphorylation inactivates PFK2, and another domain on this protein becomes active as fructose bisphosphatase-2, which converts F2,6BP back to F6P. Both glucagon and epinephrine cause high levels of cAMP in the liver. The result of lower levels of liver fructose-2,6-bisphosphate is a decrease in activity of phosphofructokinase and an increase in activity of fructose 1,6-bisphosphatase, so that gluconeogenesis (in essence, "glycolysis in reverse") is favored. This is consistent with the role of the liver in such situations, since the response of the liver to these hormones is to release glucose to the blood.

ATP competes with AMP for the allosteric effector site on the PFK enzyme. ATP concentrations in cells are much higher than those of AMP, typically 100-fold higher, but the concentration of ATP does not change more than about 10% under physiological conditions, whereas a 10% drop in ATP results in a 6-fold increase in AMP. Thus, the relevance of ATP as an allosteric effector is questionable. An increase in AMP is a consequence of a decrease in energy charge in the cell.

Citrate inhibits phosphofructokinase when tested *in vitro* by enhancing the inhibitory effect of ATP. However, it is doubtful that this is a meaningful effect *in vivo*, because citrate in the cytosol is utilized mainly for conversion to acetyl-CoA for fatty acid and cholesterol synthesis.

Pyruvate Kinase

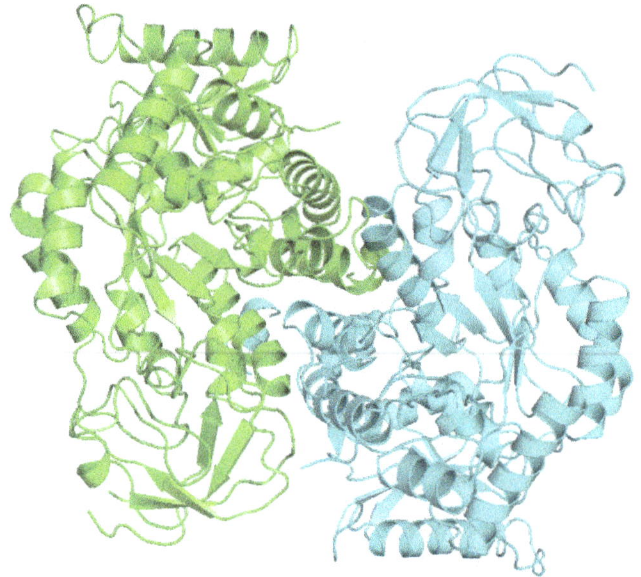

Yeast pyruvate kinase (PDB: 1A3W)

Pyruvate kinase enzyme catalyzes the last step of glycolysis, in which pyruvate and ATP are formed. Pyruvate kinase catalyzes the transfer of a phosphate group from phosphoenolpyruvate (PEP) to ADP, yielding one molecule of pyruvate and one molecule of ATP.

Liver pyruvate kinase is indirectly regulated by epinephrine and glucagon, through protein kinase A. This protein kinase phosphorylates liver pyruvate kinase to deactivate it. Muscle pyruvate kinase is not inhibited by epinephrine activation of protein kinase A. Glucagon signals fasting (no glucose available). Thus, glycolysis is inhibited in the liver but unaffected in muscle when fasting. An increase in blood sugar leads to secretion of insulin, which activates phosphoprotein phosphatase I, leading to dephosphorylation and activation of pyruvate kinase. These controls prevent pyruvate kinase from being active at the same time as the enzymes that catalyze the reverse reaction (pyruvate carboxylase and phosphoenolpyruvate carboxykinase), preventing a futile cycle.

Post-glycolysis Processes

The overall process of glycolysis is:

$$\text{Glucose} + 2\ \text{NAD}^+ + 2\ \text{ADP} + 2\ \text{P}_i \rightarrow 2\ \text{Pyruvate} + 2\ \text{NADH} + 2\ \text{H}^+ + 2\ \text{ATP} + 2\ \text{H}_2\text{O}$$

If glycolysis were to continue indefinitely, all of the NAD^+ would be used up, and glycolysis would stop. To allow glycolysis to continue, organisms must be able to oxidize NADH back to NAD^+. How this is performed depends on which external electron acceptor is available.

Anoxic Regeneration of NAD^+

One method of doing this is to simply have the pyruvate do the oxidation; in this process, pyruvate is converted to lactate (the conjugate base of lactic acid) in a process called lactic acid fermentation:

$$\text{Pyruvate} + \text{NADH} + \text{H}^+ \rightarrow \text{Lactate} + \text{NAD}^+$$

This process occurs in the bacteria involved in making yogurt (the lactic acid causes the milk to curdle). This process also occurs in animals under hypoxic (or partially anaerobic) conditions, found, for example, in overworked muscles that are starved of oxygen. In many tissues, this is a cellular last resort for energy; most animal tissue cannot tolerate anaerobic conditions for an extended period of time.

Some organisms, such as yeast, convert NADH back to NAD^+ in a process called ethanol fermentation. In this process, the pyruvate is converted first to acetaldehyde and carbon dioxide, and then to ethanol.

Lactic acid fermentation and ethanol fermentation can occur in the absence of oxygen. This anaerobic fermentation allows many single-cell organisms to use glycolysis as their only energy source.

Anoxic regeneration of NAD^+ is only an effective means of energy production during short, intense exercise in vertebrates, for a period ranging from 10 seconds to 2 minutes during a maximal effort in humans. (At lower exercise intensities it can sustain muscle activity in diving animals, such as seals, whales and other aquatic vertebrates, for very much longer periods of time.) Under these conditions NAD^+ is replenished by NADH donating its electrons to pyruvate to form lactate. This produces 2 ATP molecules per glucose molecule, or about 5% of glucose's energy potential (38 ATP molecules in bacteria). But the speed at which ATP is produced in this manner is about 100 times that of oxidative phosphorylation. The pH in the cytoplasm quickly drops when hydrogen ions accumulate in the muscle, eventually inhibiting the enzymes involved in glycolysis.

The burning sensation in muscles during hard exercise can be attributed to the production of hydrogen ions during the shift to glucose fermentation from glucose oxidation to carbon dioxide and water, when aerobic metabolism can no longer keep pace with the energy demands of the muscles. These hydrogen ions form a part of lactic acid. The body falls back on this less efficient but faster method of producing ATP under low oxygen conditions. This is thought to have been the primary means of energy production in earlier organisms before oxygen reached high concentrations in the atmosphere between 2000 and 2500 million years ago (see diagram), and thus would represent a more ancient form of energy production than the aerobic replenishment of NAD^+ in cells.

The liver in mammals gets rid of this excess lactate by transforming it back into pyruvate under aerobic conditions.

Fermenation of pyruvate to lactate is sometimes also called "anaerobic glycolysis", however, glycolysis ends with the production of pyruvate regardless of the presence or absence of oxygen.

In the above two examples of fermentation, NADH is oxidized by transferring two electrons to pyruvate. However, anaerobic bacteria use a wide variety of compounds as the terminal electron acceptors in cellular respiration: nitrogenous compounds, such as nitrates and nitrites; sulfur compounds, such as sulfates, sulfites, sulfur dioxide, and elemental sulfur; carbon dioxide; iron compounds; manganese compounds; cobalt compounds; and uranium compounds.

Aerobic Regeneration of NAD^+, and Disposal of Pyruvate

In aerobic organisms, a complex mechanism has been developed to use the oxygen in air as the final electron acceptor.

- Firstly, the NADH + H^+ generated by glycolysis has to be transferred to the mitochondrion to be oxidized, and thus to regenerate the NAD^+ necessary for glycolysis to continue. However the inner mitochondrial membrane is impermeable to NADH and NAD^+. Use is therefore made of two "shuttles" to transport the electrons from NADH across the mitochondrial membrane. They are the malate-aspartate shuttle and the glycerol phosphate shuttle. In the former the electrons from NADH are transferred to cytosolic oxaloacetate to form malate. The malate then traverses the inner mitochondrial membrane into the mitochondrial matrix, where it is reoxidized by NAD^+ forming intra-mitochondrial oxaloacetate and NADH. The oxaloacetate is then re-cycled to the cytosol via its conversion to aspartate which is readily transported out of the mitochondrion. In the glycerol phosphate shuttle electrons from cytosolic NADH are transferred to dihydroxyacetone to form glycerol-3-phosphate which readily traverses the outer mitochondrial membrane. Glycerol-3-phosphate is then reoxidized to dihydroxyacetone, donating its electrons to FAD instead of NAD^+. This reaction takes place on the inner mitochondrial membrane, allowing $FADH_2$ to donate its electrons directly to coenzyme Q (ubiquinone) which is part of the electron transport chain which ultimately transfers electrons to molecular oxygen (O_2), with the formation of water, and the release of energy eventually captured in the form of ATP.

- The glycolytic end-product, pyruvate (plus NAD^+) is converted to acetyl-CoA, CO_2 and NADH + H^+ within the mitochondria in a process called pyruvate decarboxylation.
- The resulting acetyl-CoA enters the citric acid cycle (or Krebs Cycle), where the acetyl group of the acetyl-CoA is converted into carbon dioxide by two decarboxylation reactions with the formation of yet more intra-mitochondrial NADH + H^+.
- The intra-mitochondrial NADH + H^+ is oxidized to NAD^+ by the electron transport chain, using oxygen as the final electron acceptor to form water. The energy released during this process is used to create a hydrogen ion (or proton) gradient across the inner membrane of the mitochondrion.
- Finally, the proton gradient is used to produce about 2.5 ATP for every NADH + H^+ oxidized in a process called oxidative phosphorylation.

Conversion of Carbohydrates Into Fatty Acids and Cholesterol

The pyruvate produced by glycolysis is an important intermediary in the conversion of carbohydrates into fatty acids and cholesterol. This occurs via the conversion of pyruvate into acetyl-CoA in the mitochondrion. However, this acetyl CoA needs to be transported into cytosol where the synthesis of fatty acids and cholesterol occurs. This cannot occur directly. To obtain cytosolic acetyl-CoA, citrate (produced by the condensation of acetyl CoA with oxaloacetate) is removed from the citric acid cycle and carried across the inner mitochondrial membrane into the cytosol. There it is cleaved by ATP citrate lyase into acetyl-CoA and oxaloacetate. The oxaloacetate is returned to mitochondrion as malate (and then back into oxaloacetate to transfer more acetyl-CoA out of the mitochondrion). The cytosolic acetyl-CoA can be carboxylated by acetyl-CoA carboxylase into malonyl CoA, the first committed step in the synthesis of fatty acids, or it can be combined with acetoacetyl-CoA to form 3-hydroxy-3-methylglutaryl-CoA (HMG-CoA) which is the rate limiting step controlling the synthesis of cholesterol. Cholesterol can be used as is, as a structural component of cellular membranes, or it can be used to synthesize the steroid hormones, bile salts, and vitamin D.

Conversion of Pyruvate Into Oxaloacetate for the Citric Acid Cycle

Pyruvate molecules produced by glycolysis are actively transported across the inner mitochondrial membrane, and into the matrix where they can either be oxidized and combined with coenzyme A to form CO_2, acetyl-CoA, and NADH, or they can be carboxylated (by pyruvate carboxylase) to form oxaloacetate. This latter reaction "fills up" the amount of oxaloacetate in the citric acid cycle, and is therefore an anaplerotic reaction (from the Greek meaning to "fill up"), increasing the cycle's capacity to metabolize acetyl-CoA when the tissue's energy needs (e.g. in heart and skeletal muscle) are suddenly increased by activity. In the citric acid cycle all the intermediates (e.g. citrate, iso-citrate, alpha-ketoglutarate, succinate, fumarate, malate and oxaloacetate) are regenerated during each turn of the cycle. Adding more of any of these intermediates to the mitochondrion therefore means that that additional amount is retained within the cycle, increasing all the other intermediates as one is converted into the other. Hence the addition of oxaloacetate greatly increases the amounts of all the citric acid intermediates, thereby increasing the cycle's capacity to metabolize acetyl CoA, converting its acetate component into CO_2 and water, with the release of enough energy to form 11 ATP and 1 GTP molecule for each additional molecule of acetyl CoA that combines with oxaloacetate in the cycle.

To cataplerotically remove oxaloacetate from the citric cycle, malate can be transported from the mitochondrion into the cytoplasm, decreasing the amount of oxaloacetate that can be regenerated. Furthermore, citric acid intermediates are constantly used to form a variety of substances such as the purines, pyrimidines and porphyrins.

Intermediates for Other Pathways

This article concentrates on the catabolic role of glycolysis with regard to converting potential chemical energy to usable chemical energy during the oxidation of glucose to pyruvate. Many of the metabolites in the glycolytic pathway are also used by anabolic pathways, and, as a consequence, flux through the pathway is critical to maintain a supply of carbon skeletons for biosynthesis.

The following metabolic pathways are all strongly reliant on glycolysis as a source of metabolites: and many more.

- Pentose phosphate pathway, which begins with the dehydrogenation of glucose-6-phosphate, the first intermediate to be produced by glycolysis, produces various pentose sugars, and NADPH for the synthesis of fatty acids and cholesterol.
- Glycogen synthesis also starts with glucose-6-phosphate at the beginning of the glycolytic pathway.
- Glycerol, for the formation of triglycerides and phospholipids, is produced from the glycolytic intermediate glyceraldehyde-3-phosphate.
- Various post-glycolytic pathways:
    - Fatty acid synthesis
    - Cholesterol synthesis

- The citric acid cycle which in turn leads to:
- Amino acid synthesis
- Nucleotide synthesis
- Tetrapyrrole synthesis

Although gluconeogenesis and glycolysis share many intermediates the one is not functionally a branch or tributary of the other. There are two regulatory steps in both pathways which, when active in the one pathway, are automatically inactive in the other. The two processes can therefore not be simultaneously active. Indeed, if both sets of reactions were highly active at the same time the net result would be the hydrolysis of four high energy phosphate bonds (two ATP and two GTP) per reaction cycle.

NAD^+ is the oxidizing agent in glycolysis, as it is in most other energy yielding metabolic reactions (e.g. beta-oxidation of fatty acids, and during the citric acid cycle). The NADH thus produced is primarily used to ultimately transfer electrons to O_2 to produce water, or, when O_2 is not available, to produced compounds such as lactate or ethanol. NADH is rarely used for synthetic processes, the notable exception being gluconeogenesis. During fatty acid and cholesterol synthesis the reducing agent is NADPH. This difference exemplifies a general principle that NADPH is consumed during biosynthetic reactions, whereas NADH is generated in energy-yielding reactions. The source of the NADPH is two-fold. When malate is oxidatively decarboxylated by "$NADP^+$-linked malic enzyme" pyruvate, CO_2 and NADPH are formed. NADPH is also formed by the pentose phosphate pathway which converts glucose into ribose, which can be used in synthesis of nucleotides and nucleic acids, or it can be catabolized to pyruvate.

Glycolysis in Disease

Genetic Diseases

Glycolytic mutations are generally rare due to importance of the metabolic pathway, this means that the majority of occurring mutations result in an inability for the cell to respire, and therefore cause the death of the cell at an early stage. However, some mutations are seen with one notable example being Pyruvate kinase deficiency, leading to chronic hemolytic anemia.

Cancer

Malignant rapidly growing tumor cells typically have glycolytic rates that are up to 200 times higher than those of their normal tissues of origin. This phenomenon was first described in 1930 by Otto Warburg and is referred to as the Warburg effect. The Warburg hypothesis claims that cancer is primarily caused by dysfunctionality in mitochondrial metabolism, rather than because of uncontrolled growth of cells. A number of theories have been advanced to explain the Warburg effect. One such theory suggests that the increased glycolysis is a normal protective process of the body and that malignant change could be primarily caused by energy metabolism.

This high glycolysis rate has important medical applications, as high aerobic glycolysis by malignant tumors is utilized clinically to diagnose and monitor treatment responses of cancers by imaging uptake of 2-^{18}F-2-deoxyglucose (FDG) (a radioactive modified hexokinase substrate) with positron emission tomography (PET).

There is ongoing research to affect mitochondrial metabolism and treat cancer by reducing glycolysis and thus starving cancerous cells in various new ways, including a ketogenic diet.

Alternative Nomenclature

Some of the metabolites in glycolysis have alternative names and nomenclature. In part, this is because some of them are common to other pathways, such as the Calvin cycle.

	This article		Alternative names	Alternative nomenclature
1	Glucose	Glc	Dextrose	
3	Fructose-6-phosphate	F6P		
4	Fructose-1,6-bisphosphate	F1,6BP	Fructose 1,6-diphosphate	FBP, FDP, F1,6DP
5	Dihydroxyacetone phosphate	DHAP	Glycerone phosphate	
6	Glyceraldehyde-3-phosphate	GADP	3-Phosphoglyceraldehyde	PGAL, G3P, GALP,GAP,TP
7	1,3-Bisphosphoglycerate	1,3BPG	Glycerate-1,3-bisphosphate, glycerate-1,3-diphosphate, 1,3-diphosphoglycerate	PGAP, BPG, DPG
8	3-Phosphoglycerate	3PG	Glycerate-3-phosphate	PGA, GP
9	2-Phosphoglycerate	2PG	Glycerate-2-phosphate	
10	Phosphoenolpyruvate	PEP		
11	Pyruvate	Pyr	Pyruvic acid	

Glycogenesis

Glycogenesis is the process of glycogen synthesis, in which glucose molecules are added to chains of glycogen for storage. This process is activated during rest periods following the Cori cycle, in the liver, and also activated by insulin in response to high glucose levels, for example after a carbohydrate-containing meal.

Steps

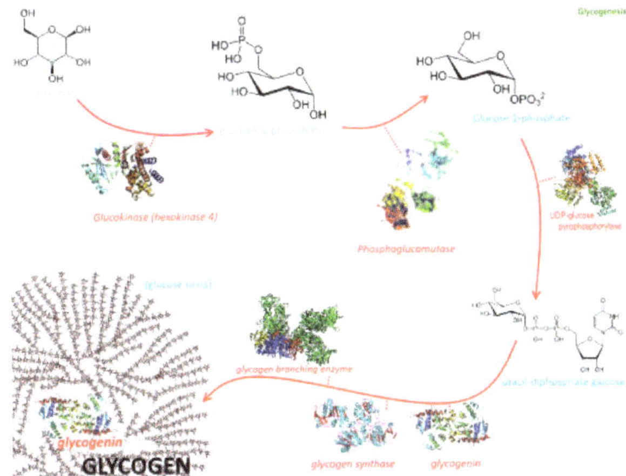

- Glucose is converted into glucose-6-phosphate by the action of glucokinase or hexokinase.
- Glucose-6-phosphate is converted into glucose-1-phosphate by the action of phosphoglucomutase, passing through the obligatory intermediate glucose-1,6-bisphosphate.
- Glucose-1-phosphate is converted into UDP-glucose by the action of the enzyme UDP-glucose phosphorylase. Pyrophosphate is formed, which is later hydrolysed by pyrophosphatase into two phosphate molecules.
- Glycogenin, a homodimer, has a tyrosine residue on each subunit that serves as the anchor for the reducing end of glycogen. Initially, about eight UDP-glucose molecules are added to each tyrosine residue by glycogenin, forming α(1→4) bonds.
- Once a chain of eight glucose monomers is formed, glycogen synthase binds to the growing glycogen chain and adds UDP-glucose to the 4-hydroxyl group of the glucosyl residue on the non-reducing end of the glycogen chain, forming more α(1→4) bonds in the process.
- Branches are made by glycogen branching enzyme (also known as amylo-α(1:4)→α(1:6)transglycosylase), which transfers the end of the chain onto an earlier part via α-1:6 glycosidic bond, forming branches, which further grow by addition of more α-1:4 glycosidic units.

Control and Regulations

Glycogenesis responds to hormonal control.

One of the main forms of control is the varied phosphorylation of glycogen synthase and glycogen phosphorylase. This is regulated by enzymes under the control of hormonal activity, which is in turn regulated by many factors. As such, there are many different possible effectors when compared to allosteric systems of regulation.

Epinephrine (Adrenaline)

Glycogen phosphorylase is activated by phosphorylation, whereas glycogen synthase is inhibited.

Glycogen phosphorylase is converted from its less active "b" form to an active "a" form by the enzyme phosphorylase kinase. This latter enzyme is itself activated by protein kinase A and deactivated by phosphoprotein phosphatase-1.

Protein kinase A itself is activated by the hormone adrenaline. Epinephrine binds to a receptor protein that activates adenylate cyclase. The latter enzyme causes the formation of cyclic AMP from ATP; two molecules of cyclic AMP bind to the regulatory subunit of protein kinase A, which activates it allowing the catalytic subunit of protein kinase A to dissociate from the assembly and to phosphorylate other proteins.

Returning to glycogen phosphorylase, the less active "b" form can itself be activated without the conformational change. 5'AMP acts as an allosteric activator, whereas ATP is an inhibitor, as already seen with phosphofructokinase control, helping to change the rate of flux in response to energy demand.

Epinephrine not only activates glycogen phosphorylase but also inhibits glycogen synthase. This amplifies the effect of activating glycogen phosphorylase. This inhibition is achieved by a similar mechanism, as protein kinase A acts to phosphorylate the enzyme, which lowers activity. This is known as co-ordinate reciprocal control. Refer to glycolysis for further information of the regulation of glycogenesis.

Insulin

Insulin has an antagonistic effect to epinephrine signaling via the beta-adrenergic receptor (G-Protein coupled receptor). When insulin binds to its receptor (insulin receptor), it results in the activation (phosphorylation) of Akt which in turn activates Phosphodiesterase (PDE). PDE then will inhibit cyclic AMP (cAMP) action and cause inactivation of PKA which will cause Hormone Sensitive Lipase (HSL) to be dephosphorylated and inactive so that lipolysis and lipogenesis is not occurring simultaneously.

Calcium Ions

Calcium ions or cyclic AMP (cAMP) act as secondary messengers. This is an example of negative control. The calcium ions activate phosphorylase kinase. This activates glycogen phosphorylase and inhibits glycogen synthase.

Allosteric Regulation

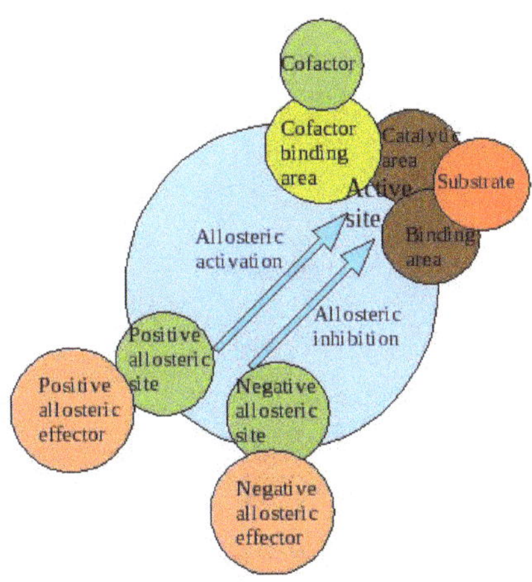

Allosteric regulation of an enzyme

In biochemistry, allosteric regulation (or allosteric control) is the regulation of a protein by binding an effector molecule at a site other than the enzyme's active site.

The site to which the effector binds is termed the *allosteric site*. Allosteric sites allow effectors to bind to the protein, often resulting in a conformational change involving protein dynamics. Effectors that enhance the protein's activity are referred to as *allosteric activators*, whereas those that decrease the protein's activity are called *allosteric inhibitors*.

Allosteric regulations are a natural example of control loops, such as feedback from downstream products or feedforward from upstream substrates. Long-range allostery is especially important in cell signaling. Allosteric regulation is also particularly important in the cell's ability to adjust enzyme activity.

The term *allostery* comes from the Greek *allos*, «other,» and *stereos*, «solid (object).» This is in reference to the fact that the regulatory site of an allosteric protein is physically distinct from its active site.

Models of Allosteric Regulation

This is a diagram of allosteric regulation of an enzyme.

Most allosteric effects can be explained by the *concerted* MWC model put forth by Monod, Wyman, and Changeux, or by the sequential model described by Koshland, Nemethy, and Filmer. Both postulate that enzyme subunits exist in one of two conformations, tensed (T) or relaxed (R), and that relaxed subunits bind substrate more readily than those in the tense state. The two models differ most in their assumptions about subunit interaction and the preexistence of both states.

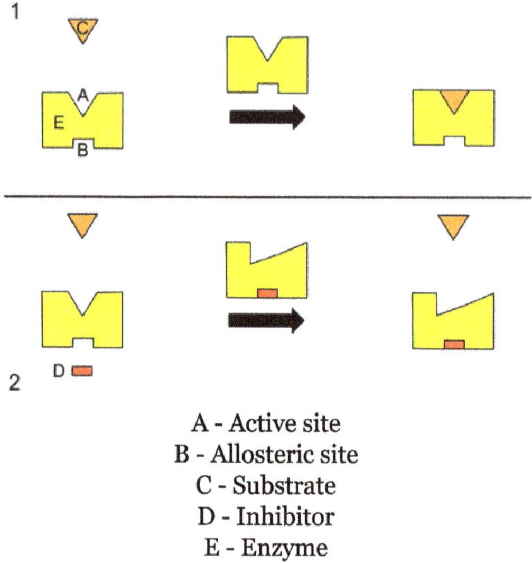

Concerted Model

The concerted model of allostery, also referred to as the symmetry model or MWC model, postulates that enzyme subunits are connected in such a way that a conformational change in one subunit is necessarily conferred to all other subunits. Thus, all subunits must exist in the same conformation. The model further holds that, in the absence of any ligand (substrate or otherwise), the equilibrium favors one of the conformational states, T or R. The equilibrium can be shifted to the R or T state through the binding of one ligand (the allosteric effector or ligand) to a site that is different from the active site (the allosteric site).

Sequential Model

The sequential model of allosteric regulation holds that subunits are not connected in such a way that a conformational change in one induces a similar change in the others. Thus, all enzyme subunits do not necessitate the same conformation. Moreover, the sequential model dictates that molecules of a substrate bind via an induced fit protocol. In general, when a subunit randomly collides with a molecule of substrate, the active site, in essence, forms a glove around its substrate. While such an induced fit converts a subunit from the tensed state to relaxed state, it does not propagate the conformational change to adjacent subunits. Instead, substrate-binding at one subunit only slightly alters the structure of other subunits so that their binding sites are more receptive to substrate. To summarize:

- subunits need not exist in the same conformation
- molecules of substrate bind via induced-fit protocol
- conformational changes are not propagated to all subunits

Morpheein Model

The morpheein model of allosteric regulation is a dissociative concerted model.

A morpheein is a homo-oligomeric structure that can exist as an ensemble of physiologically significant and functionally different alternate quaternary assemblies. Transitions between alternate morpheein assemblies involve oligomer dissociation, conformational change in the dissociated state, and reassembly to a different oligomer. The required oligomer disassembly step differentiates the morpheein model for allosteric regulation from the classic MWC and KNF models. Porphobilinogen synthase (PBGS) is the prototype morpheein.

Allosteric Resources Online

Allosteric Database

Allostery is a direct and efficient means for regulation of biological macromolecule function, produced by the binding of a ligand at an allosteric site topographically distinct from the orthosteric site. Due to the often high receptor selectivity and lower target-based toxicity, allosteric regulation is also expected to play an increasing role in drug discovery and bioengineering. The AlloSteric Database (ASD, http://mdl.shsmu.edu.cn/ASD) provides a central resource for the display, search and analysis of the structure, function and related annotation for allosteric molecules. Currently, ASD contains allosteric proteins from more than 100 species and modulators in three categories (activators, inhibitors, and regulators). Each protein is annotated with detailed description of allostery, biological process and related diseases, and each modulator with binding affinity, physicochemical properties and therapeutic area. Integrating the information of allosteric proteins in ASD should allow the prediction of allostery for unknown proteins, to be followed with experimental validation. In addition, modulators curated in ASD can be used to investigate potential allosteric targets for a query compound, and can help chemists to implement structure modifications for novel allosteric drug design.

Allosteric Residues & their Prediction Using the StRESS Web Server

Not all protein residues play equally important roles in allosteric regulation. The identification of residues that are essential to allostery (so-called "allosteric residues") has been the focus of many studies, especially within the last decade. In part, this growing interest is a result of their general importance in protein science, but also because allosteric residues may be exploited in biomedical contexts. Pharmacologically important proteins with difficult-to-target sites may yield to approaches in which one alternatively targets easier-to-reach residues that are capable of allosterically regulating the primary site of interest. These residues can broadly be classified as surface- and interior-allosteric amino acids. Allosteric sites at the surface generally play regulatory roles that are fundamentally distinct from those within the interior; surface residues may serve as receptors or effector sites in allosteric signal transmission, whereas those within the interior may act to transmit such signals. STRucturally-identified ESSential residues (STRESS, http://stress.molmovdb.org) is a web tool that enables users to submit their own protein structures of interest in order to predict both surface- and interior-allosteric residues in an algorithmically efficient manner. The software behind this server employs 3D structures to build models of conformational change in order to perform predictions.

Allosteric Modulation

Positive Modulation

Positive allosteric modulation (also known as *allosteric activation*) occurs when the binding of one ligand enhances the attraction between substrate molecules and other binding sites. An example is the binding of oxygen molecules to hemoglobin, where oxygen is effectively both the substrate and the effector. The allosteric, or "other", site is the active site of an adjoining protein subunit. The binding of oxygen to one subunit induces a conformational change in that subunit that interacts with the remaining active sites to enhance *their* oxygen affinity. Another example of allosteric activation is seen in cytosolic IMP-GMP specific 5'-nucleotidase II (cN-II), where the affinity for substrate GMP increases upon GTP binding at the dimer interface

Negative Modulation

Negative allosteric modulation (also known as *allosteric inhibition*) occurs when the binding of one ligand decreases the affinity for substrate at other active sites. For example, when 2,3-BPG binds to an allosteric site on hemoglobin, the affinity for oxygen of all subunits decreases. This is when a regulator is absent from the binding site.

Direct thrombin inhibitors provides an excellent example of negative allosteric modulation. Allosteric inhibitors of thrombin have been discovered which could potentially be used as anticoagulants.

Another example is strychnine, a convulsant poison, which acts as an allosteric inhibitor of the glycine receptor. Glycine is a major post-synaptic inhibitory neurotransmitter in mammalian spinal cord and brain stem. Strychnine acts at a separate binding site on the glycine receptor in an allosteric manner; i.e., its binding lowers the affinity of the glycine receptor for glycine. Thus, strychnine inhibits the action of an inhibitory transmitter, leading to convulsions.

Another instance in which negative allosteric modulation can be seen is between ATP and the enzyme Phosphofructokinase within the negative feedback loop that regulates glycolysis. Phosphofructokinase (generally referred to as PFK) is an enzyme that catalyses the third step of glycolysis: the phosphorylation of Fructose-6-phosphate into Fructose 1,6-bisphosphate. PFK can be allosterically inhibited by high levels of ATP within the cell. When ATP levels are high, ATP will bind to an allosteoric site on phosphofructokinase, causing a change in the enzyme's three-dimensional shape. This change causes its affinity for substrate (fructose-6-phosphate and ATP) at the active site to decrease, and the enzyme is deemed inactive. This causes glycolysis to cease when ATP levels are high, thus conserving the body's glucose and maintaining balanced levels of cellular ATP. In this way, ATP serves as a negative allosteric modulator for PFK, despite the fact that it is also a substrate of the enzyme.

Types of Allosteric Regulation

Homotropic

A homotropic allosteric modulator is a substrate for its target enzyme, as well as a regulatory molecule of the enzyme's activity. It is typically an activator of the enzyme. For example, O_2 and CO are homotropic allosteric modulators of hemoglobin.

Heterotropic

A heterotropic allosteric modulator is a regulatory molecule that is not the enzyme's substrate. It may be either an activator or an inhibitor of the enzyme. For example, H^+, CO_2, and 2,3-bisphosphoglycerate are heterotropic allosteric modulators of hemoglobin.

Some allosteric proteins can be regulated by both their substrates and other molecules. Such proteins are capable of both homotropic and heterotropic interactions.

Non-regulatory Allostery

A non-regulatory allosteric site refers to any non-regulatory component of an enzyme (or any protein), that is not itself an amino acid. For instance, many enzymes require sodium binding to ensure proper function. However, the sodium does not necessarily act as a regulatory subunit; the sodium is always present and there are no known biological processes to add/remove sodium to regulate enzyme activity. Non-regulatory allostery could comprise any other ions besides sodium (calcium, magnesium, zinc), as well as other chemicals and possibly vitamins.

Pharmacology

Allosteric modulation of a receptor results from the binding of allosteric modulators at a different site (a "regulatory site") from that of the endogenous ligand (an "active site") and enhances or inhibits the effects of the endogenous ligand. Under normal circumstances, it acts by causing a conformational change in a receptor molecule, which results in a change in the binding affinity of the ligand. In this way, an allosteric ligand modulates the receptor's activation by its primary (orthosteric) ligand, and can be thought to act like a dimmer switch in an electrical circuit, adjusting the intensity of the response.

For example, the $GABA_A$ receptor has two active sites that the neurotransmitter gamma-aminobutyric acid (GABA) binds, but also has benzodiazepine and general anaesthetic agent regulatory binding sites. These regulatory sites can each produce positive allosteric modulation, potentiating the activity of GABA. Diazepam is an agonist at the benzodiazepine regulatory site, and its antidote flumazenil is an antagonist.

More recent examples of drugs that allosterically modulate their targets include the calcium-mimicking cinacalcet and the HIV treatment maraviroc.

Allosteric Sites as Drug Targets

Allosteric sites may represent a novel drug target. There are a number of advantages in using allosteric modulators as preferred therapeutic agents over classic orthosteric ligands. For example, G protein-coupled receptor (GPCR) allosteric binding sites have not faced the same evolutionary pressure as orthosteric sites to accommodate an endogenous ligand, so are more diverse. Therefore, greater GPCR selectivity may be obtained by targeting allosteric sites. This is particularly useful for GPCRs where selective orthosteric therapy has been difficult because of sequence conservation of the orthosteric site across receptor subtypes. Also, these modulators have a decreased potential for toxic effects, since modulators with limited co-operativity will have a ceiling level to their effect, irrespective of the administered dose. Another type of pharmacological selectivity that is unique to allosteric modulators is based on co-operativity. An allosteric modulator may display neutral co-operativity with an orthosteric ligand at all subtypes of a given receptor except the subtype of interest, which is termed "absolute subtype selectivity". If an allosteric modulator does not possess appreciable efficacy, it can provide another powerful therapeutic advantage over orthosteric ligands, namely the ability to selectively tune up or down tissue responses only when the endogenous agonist is present. Oligomer-specific small molecule binding sites are drug targets for medically relevant morpheeins.

Phosphofructokinase 1

Phosphofructokinase-1 (PFK-1) is one of the most important regulatory enzymes (EC 2.7.1.11) of glycolysis. It is an allosteric enzyme made of 4 subunits and controlled by many activators and inhibitors. PFK-1 catalyzes the important "committed" step of glycolysis, the conversion of fructose 6-phosphate and ATP to fructose 1,6-bisphosphate and ADP. Glycolysis is the foundation for respiration, both anaerobic and aerobic. Because phosphofructokinase (PFK) catalyzes the ATP-dependent phosphorylation to convert fructose-6-phosphate into fructose 1,6-bisphosphate and ADP, it is one of the key regulatory and rate limiting steps of glycolysis. PFK is able to regulate glycolysis through allosteric inhibition, and in this way, the cell can increase or decrease the rate of glycolysis in response to the cell's energy requirements. For example, a high ratio of ATP to ADP will inhibit PFK and glycolysis. The key difference between the regulation of PFK in eukaryotes and prokaryotes is that in eukaryotes PFK is activated by fructose 2,6-bisphosphate. The purpose of fructose 2,6-bisphosphate is to supersede ATP inhibition, thus allowing eukaryotes to have greater sensitivity to regulation by hormones like glucagon and insulin.

Structure

Mammalian PFK1 is a 340kd tetramer composed of different combinations of three types of subunits: muscle (M), liver (L), and platelet (P). The composition of the PFK1 tetramer differs according to the tissue type it is present in. For example, mature muscle expresses only the M isozyme, therefore, the muscle PFK1 is composed solely of homotetramers of M4. The liver and kidneys express predominantly the L isoform. In erythrocytes, both M and L subunits randomly tetramerize to form M4, L4 and the three hybrid forms of the enzyme (ML3, M2L2, M3L). As a result, the kinetic and regulatory properties of the various isoenzymes pools are dependent on subunit composition. Tissue-specific changes in PFK activity and isoenzymic content contribute significantly to the diversities of glycolytic and gluconeogenic rates which have been observed for different tissues.

PFK1 is an allosteric enzyme and has a structure similar to that of hemoglobin in so far as it is a dimer of a dimer. One half of each dimer contains the ATP binding site whereas the other half the substrate (fructose-6-phosphate or (F6P)) binding site as well as a separate allosteric binding site.

Each subunit of the tetramer is 319 amino acids and consists of two domain, one that binds the substrate ATP, and the other that binds fructose-6-phosphate. Each domain is a b barrel, and has cylindrical b sheet surrounded by alpha helices.

On the opposite side of the each subunit from each active site is the allosteric site, at the interface between subunits in the dimer. ATP and AMP compete for this site. The N-terminal domain has a catalytic role binding the ATP, and the C-terminal has a regulatory role

Mechanism

PFK1 is an allosteric enzyme whose activity can be described using the symmetry model of allosterism whereby there is a concerted transition from an enzymatically inactive T-state to the active R-state. F6P binds with a high affinity to the R state but not the T state enzyme. For every molecule of F6P that binds to PFK1, the enzyme progressively shifts from T state to the R state. Thus a graph plotting PFK1 activity against increasing F6P concentrations would adopt the sigmoidal curve shape traditionally associated with allosteric enzymes.

PFK1 belongs to the family of phosphotransferases and it catalyzes the transfer of γ-phosphate from ATP to fructose-6-phosphate. The PFK1 active site comprises both the ATP-Mg2+ and the F6P binding sites. Some proposed residues involved with substrate binding in *E. coli* PFK1 include Asp127 and Arg171. In B. stearothermophilus PFK1, the positively charged side chain of Arg162 residue forms a hydrogen-bonded salt bridge with the negatively charged phosphate group of F6P, an interaction which stabilizes the R state relative to the T state and is partly responsible for the homotropic effect of F6P binding. In the T state, enzyme conformation shifts slightly such that the space previously taken up by the Arg162 is replaced with Glu161. This swap in positions between adjacent amino acid residues inhibits the ability of F6P to bind the enzyme.

Allosteric activators such as AMP and ADP bind to the allosteric site as to facilitate the formation of the R state by inducing structural changes in the enzyme. Similarly, inhibitors such as ATP and PEP bind to the same allosteric site and facilitate the formation of the T state, thereby inhibiting enzyme activity.

The hydroxyl oxygen of carbon 1 does a nucleophilic attack on the beta phosphate of ATP. These electrons are pushed to the anhydride oxygen between the beta and gamma phosphates of ATP.

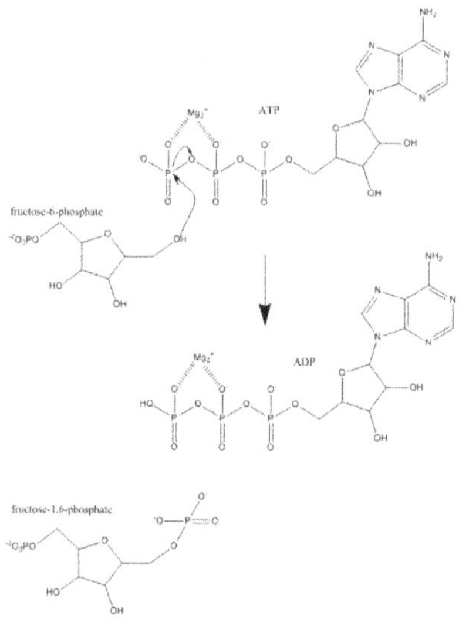

Mechanism of phosphofructokinase 1

Regulation

PFK1 is the most important control site in the mammalian glycolytic pathway. This step is subject to extensive regulation since it is not only highly exergonic under physiological conditions, but also because it is a committed step – the first irreversible reaction unique to the glycolytic pathway. This leads to a precise control of glucose and the other monosaccharides galactose and fructose going down the glycolytic pathway. Before this enzyme's reaction, glucose-6-phosphate can potentially travel down the pentose phosphate pathway, or be converted to glucose-1-phosphate for glycogenesis.

PFK1 is allosterically inhibited by high levels of ATP but AMP reverses the inhibitory action of ATP. Therefore, the activity of the enzyme increases when the cellular ATP/AMP ratio is lowered. Glycolysis is thus stimulated when energy charge falls. PFK1 has two sites with different affinities for ATP which is both a substrate and an inhibitor.

PFK1 is also inhibited by low pH levels which augment the inhibitory effect of ATP. The pH falls when muscle is functioning anaerobically and producing excessive quantities of lactic acid (although lactic acid is not itself the cause of the decrease in pH). This inhibitory effect serves to protect the muscle from damage that would result from the accumulation of too much acid.

Finally, PFK1 is allosterically inhibited by PEP, citrate, and ATP. Phosphoenolpyruvic acid is a product further downstream the glycolytic pathway. Although citrate does build up when the Krebs Cycle enzymes approach their maximum velocity, it is questionable whether citrate accumulates to a sufficient concentration to inhibit PFK-1 under normal physiological conditions. ATP concentration build up indicates an excess of energy and does have an allosteric modulation site on PFK1 where it decreases the affinity of PFK1 for its substrate.

PFK1 is allosterically activated by a high concentration of AMP, but the most potent activator is fructose 2,6-bisphosphate, which is also produced from fructose-6-phosphate by PFK2. Hence, an abundance of F6P results in a higher concentration of fructose 2,6-bisphosphate (F-2,6-BP). The binding of F-2,6-BP increases the affinity of PFK1 for F6P and diminishes the inhibitory effect of ATP. This is an example of feedforward stimulation as glycolysis is accelerated when glucose is abundant.

PFK is inhibited by glucagon through repression of synthesis. Glucagon activates protein kinase A which, in turn, shuts off the kinase activity of PFK2. This reverses any synthesis of F-2,6-BP from F6P and thus inhibits PFK1 activity.

The precise regulation of PFK1 prevents glycolysis and gluconeogenesis from occurring simultaneously. However, there is substrate cycling between F6P and F-1,6-BP. Fructose-1,6-bisphosphatase (FBPase) catalyzes the hydrolysis of F-1,6-BP back to F6P, the reverse reaction catalyzed by PFK1. There is a small amount of FBPase activity during glycolysis and some PFK1 activity during gluconeogenesis. This cycle allows for the amplification of metabolic signals as well as the generation of heat by ATP hydrolysis.

Serotonin (5-HT) increases PFK by binding to the 5-HT(2A) receptor, causing the tyrosine residue of PFK to be phosphorylated via phospholipase C. This in turn redistributes PFK within the skeletal muscle cells. Because PFK regulates glycolytic flux, serotonin plays a regulatory role in glycolysis

Genes

There are three phosphofructokinase genes in humans:

- PFKL – liver
- PFKM – muscle
- PFKP – platelet

Clinical Significance

A genetic mutation in the PFKM gene results in Tarui's disease, which is a glycogen storage disease where the ability of certain cell types to utilize carbohydrates as a source of energy is impaired.

Tarui disease is a glycogen storage disease with symptoms including muscle weakness (myopathy) and exercise induced cramping and spasms, myoglobinuria (presence of myoglobin in urine, indicating muscle destruction) and compensated hemolysis. ATP is a natural allosteric inhibitor of PFK, in order to prevent unnecessary production of ATP through glycolysis. However, a mutation in Asp(543)Ala can result in ATP having a stronger inhibitory effect (due to increased binding to PFK's inhibitory allosteric binding site).

Phosphofructokinase mutation and cancer: In order for cancer cells to meet their energy requirements due to their rapid cell growth and division, they survive more effectively when they have a hyperactive phosphofructokinase 1 enzyme. When cancer cells grow and divide quickly, they initially do not have as much blood supply, and can thus have hypoxia (oxygen deprivation), and this triggers O-GlcNAcylation at serine 529 of PFK, giving a selective growth advantage to cancer cells.

Herpes simplex type 1 and phosphofructokinase: Some viruses, including HIV, HCMV, Mayaro, and HCMV affect cellular metabolic pathways such as glycolysis by a MOI-dependent increase in the activity of PFK. The mechanism that Herpes increases PFK activity is by phosphorylating the enzyme at the serine residues. The HSV-1 induced glycolysis increases ATP content, which is critical for the virus's replication.

Pyruvate Kinase

Pyruvate kinase is the enzyme that catalyzes the final step of glycolysis. It catalyzes the transfer of a phosphate group from phosphoenolpyruvate (PEP) to adenosine diphosphate (ADP), yielding one molecule of pyruvate and one molecule of ATP. Pyruvate kinase is present in four distinct, tissue-specific isozymes, each consisting of particular kinetic properties necessary to accommodate the variations in metabolic requirements of diverse tissues.

Isozymes in Vertebrates

There are four isozymes of pyruvate kinase in vertebrates: L (liver), R (erythrocytes), M1(muscles, hearts and brain) and M2 (only form detectable in early fetal tissue and present in most adult tissues). R and L isozymes differ from M1 and M2 in that they are both exclusively allosterically and reversibly regulated. From a kinetic standpoint, the R and L isozymes of pyruvate kinase have two key conformation states; one with a high substrate affinity and one with a low substrate affinity. The R-state, characterized by high substrate affinity, serves as the activated form of pyruvate kinase and is stabilized by PEP and FBP, promoting the glycolytic pathway. The T-state, characterized by low substrate affinity, serves as the inactivated form of pyruvate kinase, bound and stabilized by ATP and alanine, causing phosphorylation of pyruvate kinase and the inhibition of glycolysis.

Gene expression varies between the different isozymes. M1 and M2 isozymes are regulated by the gene PKM and R and L isozymes are regulated by the gene PKLR. In terms of structure, there is both a tetrameric and dimeric form of pyruvate kinase. The tetrameric form is the pyruvate kinase structure in its R-state conformation, namely with high binding affinity to PEP. In contrast, the dimeric form is its structure in T-state conformation, namely with a low binding affinity to PEP. As a result, gene expression can be regulated by converting the highly active tetrameric form of PKM2, which yields high PEP concentrations, into an inactive dimeric form, which yields a PEP concentration of nearly zero.

Reaction

Glycolysis

There are two steps in the pyruvate kinase reaction in glycolysis. First, PEP transfers a phosphate group to ADP, producing ATP and the enolate of pyruvate. Secondly, a proton must be added to the enolate of pyruvate to produce the functional form of pyruvate that the cell requires.

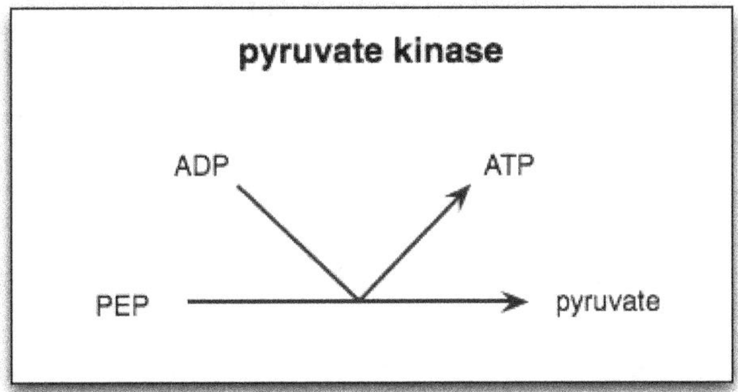

A simple diagram demonstrating the final step of glycolysis, the transfer of a phosphate group from phosphoenolpyruvate (PEP) to adenosine diphosphate (ADP) by pyruvate kinase, yielding one molecule of pyruvate and one molecule of ATP.

In yeast cells, the interaction of yeast pyruvate kinase (YPK) with PEP and its allosteric effector Fructose 1,6-bisphosphate (FBP,) was found to be enhanced by the presence of Mg. Therefore, Mg was isolated as an important component in the successful catalysis of PEP into pyruvate by pyruvate kinase. Furthermore, the metal ion Mn^{2+} was shown to have a similar, but stronger effect on the coupling free energy of YPK than Mg. The binding of metal ions to the metal binding sites on pyruvate kinase enhance the rate of this glycolytic reaction.

The glycolytic reaction catalyzed by pyruvate kinase is the final step of glycolysis. It is one of the three rate-affecting steps of the catabolic reaction cascade. The rate-affecting steps are the slower steps of a reaction and thus determines the rate of the overall reaction. In glycolysis, the rate-affecting steps are coupled with the hydrolysis of ATP or the phosphorylation of ADP to create the highly energetically favorable and irreversible reaction mechanism. This final step is highly regulated and deliberately irreversible because pyruvate is a crucial intermediate building block for further metabolic pathways. Once pyruvate kinase synthesizes pyruvate, pyruvate either enters the TCA cycle for further production of ATP under aerobic conditions, or is reduced to lactate under anaerobic conditions. Both of these secondary metabolic pathways are essential to the function of the metabolism.

Gluconeogenesis: the Reverse Reaction

Pyruvate kinase also serves as a regulatory enzyme for gluconeogenesis, a biochemical pathway in which the liver generates glucose from pyruvate and other substrates. Gluconeogenesis utilizes noncarbohydrate sources to provide glucose to the brain and red blood cells in times of starvation when direct glucose reserves are exhausted. During a fasting state, pyruvate carboxylase phosphorylates pyruvate kinase, inactivating the enzyme and therefore, preventing phosphoenolpyruvate from being converted into pyruvate. Instead, phosphoenolpyruvate is converted into glucose via a cascade of gluconeogenesis reactions. Although it utilizes similar enzymes, gluconeogenesis is not the reverse of glycolysis. It is instead a pathway that circumvents the irreversible steps of glycolysis. Furthermore, gluconeogenesis and glycolysis do not occur concurrently in the cell at any given moment as they are reciprocally regulated by cell signaling. Once the gluconeogenesis pathway is complete, the glucose produced is expelled from the liver, proving energy for the vital tissues in the fasting state.

Regulation

Glycolysis is highly regulated at three of its catalytic steps: the phosphorylation of glucose by hexokinase, the phosphorylation of fructose-6-phosphate by phosphofructokinase, and the transfer of phosphate from PEP to ADP by pyruvate kinase. Under wild-type conditions, all three of these reactions are irreversible, have a large negative free energy and are responsible for the regulation of this pathway. Pyruvate kinase activity is most broadly regulated by allosteric effectors, covalent modifiers and hormonal control. However, the most significant pyruvate kinase regulator is fructose-1,6-bisphosphate (FBP), which serves as an allosteric effector for the enzyme.

Allosteric Effectors

Allosteric regulation is the binding of an effector to a site on the protein other than the active site, causing a conformational change and altering the activity of that given protein or enzyme. Pyruvate kinase has been found to be allosterically activated by FBP and allosterically inactivated by ATP and alanine.

Fructose-1,6-bisphosphate

FBP is the most significant source of regulation because it comes from within the glycolysis pathway. FBP is a glycolytic intermediate produced from the phosphorylation of fructose 6-phosphate. FBP binds to the allosteric binding site on domain C of pyruvate kinase and changes the conformation of the enzyme, causing the activation of pyruvate kinase activity. As an intermediate present within the glycolytic pathway, FBP provides feedforward stimulation because the higher the concentration of FBP, the greater the allosteric activation and magnitude of pyruvate kinase activity. Pyruvate kinase is most sensitive to the effects of FBP. As a result, the remainder of the regulatory mechanisms serve as secondary modification.

Covalent Modifiers

Covalent modifiers serve as indirect regulators by controlling the phosphorylation and dephosphorylation of enzymes, resulting in the activation and inhibition of enzymatic activity. In the liver, glucagon and epinephrine serve as covalent modifiers by activating protein kinase A which in turn phosphorylates, and deactivates pyruvate kinase. In contrast, the secretion of insulin in response to blood sugar elevation activates phosphoprotein phosphatase I, causing the dephosphorylation and activation of pyruvate kinase. This regulation system is responsible for the avoidance of a futile cycle through the prevention of simultaneous activation of pyruvate kinase and enzymes that catalyze gluconeogenesis.

Carbohydrate Response Element Binding Protein (ChREBP)

ChREBP is found to be an essential protein in gene transcription of the L isozyme of pyruvate kinase. The domains of ChREBP are target sites for regulation of pyruvate kinase by glucose and cAMP. Specifically, ChREBP is activated by a high concentration of glucose and inhibited by cAMP. Glucose and cAMP work in opposition with one another through covalent modifier regulation. While cAMP binds to Ser196 and Thr666 binding sites of ChREBP, causing the phosphorylation and inactivation of pyruvate kinase; glucose binds to Ser196 and Thr666 binding sites of ChREBP, causing the dephosphorylation and activation of pyruvate kinase. As a result, cAMP and excess carbohydrates are shown to play an indirect role in pyruvate kinase regulation.

Hormonal Control

In order to prevent a futile cycle, glycolysis and gluconeogenesis are heavily regulated in order to ensure that they are never operating in the cell at the same time. As a result, the inhibition of pyruvate kinase by glucagon, cyclic AMP and epinephrine, not only shuts down glycolysis, but also stimulates gluconeogenesis. Alternatively, insulin interferes with the effect of glucagon, cyclic AMP and epinephrine, causing pyruvate kinase to function normally and gluconeogenesis to be shut down. Furthermore, glucose was found to inhibit and disrupt gluconeogenesis, leaving pyruvate kinase activity and glycolysis unaffected. Overall, the interaction between hormones plays a key role in the functioning and regulation of glycolysis and gluconeogenesis in the cell.

Inhibitory Effect of Metformin

Metformin, or dimethylbiguanide, is the primary treatment used for type 2 diabetes. Metaformin has been shown to indirectly affect pyruvate kinase through the inhibition of gluconeogenesis. Specifically, the addition of metformin is linked to a marked decrease in glucose flux and increase in lactate/pyruvate flux from various metabolic pathways. Although metaformin does not directly affect pyruvate kinase activity, it causes a decrease in the concentration of ATP. Due to the allosteric inhibitory effects of ATP on pyruvate kinase, a decrease in ATP results in diminished inhibition and the subsequent stimulation of pyruvate kinase. Consequently, the increase in pyruvate kinase activity directs metabolic flux through glycolysis rather than gluconeogenesis.

Clinical Applications

Deficiency

Genetic defects of this enzyme cause the disease known as pyruvate kinase deficiency. In this condition, a lack of pyruvate kinase slows down the process of glycolysis. This effect is especially devastating in cells that lack mitochondria, because these cells must use anaerobic glycolysis as their sole source of energy because the TCA cycle is not available. For example, red blood cells, which in a state of pyruvate kinase deficiency, rapidly become deficient in ATP and can undergo hemolysis. Therefore, pyruvate kinase deficiency can cause chronic nonspherocytic hemolytic anemia (CNSHA).

PK-LR Gene Mutation

Pyruvate kinase deficiency is caused by an autosomal recessive trait. Mammals have two pyruvate kinase genes, PK-LR (which encodes for pyruvate kinase isozymes L and R) and PK-M (which encodes for pyruvate kinase isozyme M1), but only PKLR encodes for the red blood isozyme which effects pyruvate kinase deficiency. Over 250 PK-LR gene mutations have been identified and associated with pyruvate kinase deficiency.

DNA testing has guided the discovery of the location of PKLR on chromosome 1 and the development of direct gene sequencing tests to molecularly diagnose pyruvate kinase deficiency.

Applications of Pyruvate Kinase Inhibition

Reactive Oxygen Species (ROS) Inhibition

Reactive oxygen species (ROS) are chemically reactive forms of oxygen. In human lung cells, ROS has been shown to inhibit the M2 isozyme of pyruvate kinase (PKM2). ROS achieves this inhibition by oxidizing Cys358 and inactivating PKM2. As a result of PKM2 inactivation, glucose flux is no longer converted into pyruvate, but is instead utilized in the pentose phosphate pathway, resulting in the reduction and detoxification of ROS. In this manner, the harmful effects of ROS are increased and cause greater oxidative stress on the lung cells, leading to potential tumor formation. This inhibitory mechanism is important because it may suggest that the regulatory mechanisms in PKM2 are responsible for aiding cancer cell resistance to oxidative stress and enhanced tumorigenesis.

Phenylalanine Inhibition

Phenylalanine is found to function as a competitive inhibitor of pyruvate kinase in the brain. Although the degree of phenylalanine inhibitory activity is similar in both fetal and adult cells, the enzymes in the fetal brain cells are significantly more vulnerable to inhibition than those in adult brain cells. A study of PKM2 in babies with the genetic brain disease phenylketonurics (PKU), showed elevated levels of phenylalanine and decreased effectiveness of PKM2. This inhibitory mechanism provides insight into the role of pyruvate kinase in brain cell damage.

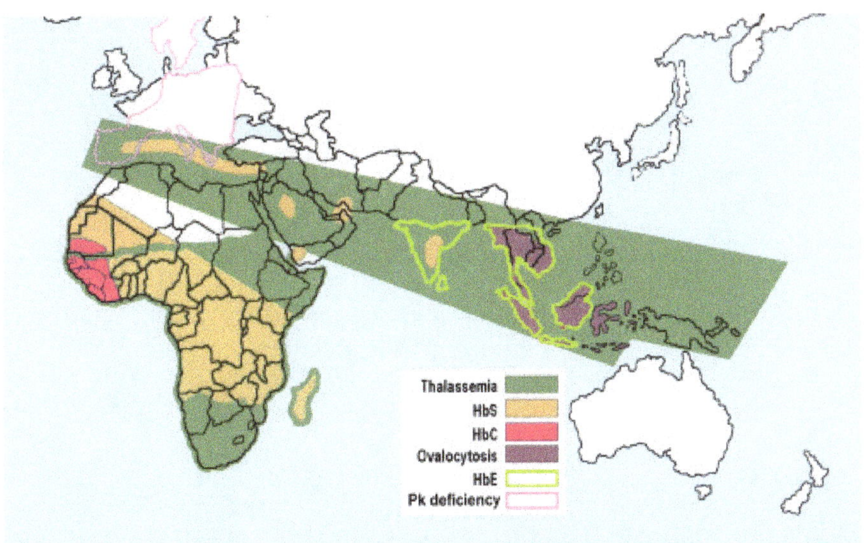

Distribution of red blood cell abnormalities worldwide

Alternatives

A reversible enzyme with a similar function, pyruvate phosphate dikinase (PPDK), is found in some bacteria and has been transferred to a number of anaerobic eukaryote groups (for example, *Streblomastix*, *Giardia*, *Entamoeba*, and *Trichomonas*), it seems via horizontal gene transfer on two or more occasions. In some cases, the same organism will have both pyruvate kinase and PPDK.

References

- Stryer, Lubert (1995). "Glycolysis.". In: Biochemistry. (Fourth ed.). New York: W.H. Freeman and Company. pp. 483–508. ISBN 0 7167 2009 4.
- Anderson, Douglas M., ed. (2003). Dorland's Illustrated Medical Dictionary (30th ed.). Philadelphia, PA: Saunders. pp. 35, 71. ISBN 0 8089 2288 2.
- Garrett, R.; Grisham, C. M. (2005). Biochemistry (3rd ed.). Belmont, CA: Thomson Brooks/Cole. pp. 582–583. ISBN 0-534-49033-6.
- Voet, Donald; Judith G. Voet; Charlotte W. Pratt (2006). Fundamentals of Biochemistry, 2nd Edition. John Wiley and Sons, Inc. pp. 547, 556. ISBN 0-471-21495-7.
- Berg, Jeremy M.; Tymoczko, John L.; Stryer, Lubert; Berg, Jeremy M.; Tymoczko, John L.; Stryer, Lubert (2002-01-01). Biochemistry (5th ed.). W H Freeman. ISBN 0716730510.
- "Embden, Gustav – Dictionary definition of Embden, Gustav | Encyclopedia.com: FREE online dictionary". www.encyclopedia.com. Retrieved 2016-02-23.

6

Fermentation in Food Processing

Fermentation can be made to occur under certain conditions to aid in the preparation of food like wine, pickles, yoghurt, stinky tofu, vinegar etc. Fermented foods are full of beneficial enzymes and various strains of probiotics that help keep the human gut functioning well. This chapter expounds about the role of fermentation in food processing, wine making and brewing. It also studies food microbiology that is concerned with the microorganisms that thrive in food and are responsible for food contamination or food spoilage.

Fermentation in Food Processing

Grapes being trodden to extract the juice and made into wine in storage jars. Tomb of Nakht, 18th dynasty, Thebes, Ancient Egypt

Fermentation in food processing is the process of converting carbohydrates to alcohol or organic acids using microorganisms—yeasts or bacteria—under anaerobic conditions. Fermentation usually implies that the action of microorganisms is desired. The science of fermentation is known as zymology or zymurgy.

The term fermentation sometimes refers specifically to the chemical conversion of sugars into ethanol, producing alcoholic drinks such as wine, beer, and cider. However, similar processes take place in the leavening of bread (CO_2 produced by yeast activity), and in the preservation of sour foods with the production of lactic acid, such as in sauerkraut and yogurt.

Apart from alcohol, widely consumed fermented foods include vinegar, olives, yogurt, bread, and cheese. In various parts of the world, more localised foods prepared by fermentation may also be based on beans, dough, grain, vegetables, fruit, honey, dairy products, fish, meat, or tea.

History and Prehistory

Natural fermentation precedes human history. Since ancient times, humans have exploited the fermentation process. The earliest evidence of an alcoholic drink, made from fruit, rice, and honey, dates from 7000 to 6600 BC, in the Neolithic Chinese village of Jiahu, and winemaking dates from 6000 BC, in Georgia, in the Caucasus area. Seven-thousand-year-old jars containing the remains of wine, now on display at the University of Pennsylvania, were excavated in the Zagros Mountains in Iran. There is strong evidence that people were fermenting alcoholic drinks in Babylon c. 3000 BC, ancient Egypt c. 3150 BC, pre-Hispanic Mexico c. 2000 BC, and Sudan c. 1500 BC.

The French chemist Louis Pasteur founded zymology, when in 1856 he connected yeast to fermentation. When studying the fermentation of sugar to alcohol by yeast, Pasteur concluded that the fermentation was catalyzed by a vital force, called "ferments", within the yeast cells. The "ferments" were thought to function only within living organisms. "Alcoholic fermentation is an act correlated with the life and organization of the yeast cells, not with the death or putrefaction of the cells", he wrote.

Nevertheless, it was known that yeast extracts can ferment sugar even in the absence of living yeast cells. While studying this process in 1897, Eduard Buchner of Humboldt University of Berlin, Germany, found that sugar was fermented even when there were no living yeast cells in the mixture, by a yeast secreted enzyme complex that he termed *zymase*. In 1907 he received the Nobel Prize in Chemistry for his research and discovery of "cell-free fermentation".

One year earlier, in 1906, ethanol fermentation studies led to the early discovery of NAD·

Uses

Beer and bread, two major uses of fermentation in food

Food fermentation is the conversion of sugars and other carbohydrates into alcohol or preservative organic acids and carbon dioxide. All three products have found human uses. The production of alcohol is made use of when fruit juices are converted to wine, when grains are made into beer, and when foods rich in starch, such as potatoes, are fermented and then distilled to make spirits such as gin and vodka. The production of carbon dioxide is used to leaven bread. The production of organic acids is exploited to preserve and flavor vegetables and dairy products.

Food fermentation serves five main purposes: to enrich the diet through development of a diversity of flavors, aromas, and textures in food substrates; to preserve substantial amounts of food through lactic acid, alcohol, acetic acid, and alkaline fermentations; to enrich food substrates with protein, essential amino acids, and vitamins; to eliminate antinutrients; and to reduce cooking time and the associated use of fuel.

Fermented Foods by Region

Nattō, a Japanese fermented soybean food

- Worldwide: alcohol(beer, wine), vinegar, olives, yogurt, bread, cheese
- Asia
    - East and Southeast Asia: amazake, atchara, bai-ming, belacan, burong mangga, com ruou, dalok, doenjang, douchi, jeruk, lambanog, kimchi, kombucha, leppet-so, narezushi, miang, miso, nata de coco, nata de pina, natto, naw-mai-dong, oncom, pak-siam-dong, paw-tsaynob, prahok, ruou nep, sake, seokbakji, soju, soy sauce, stinky tofu, szechwan cabbage, tai-tan tsoi, chiraki, tape, tempeh, totkal kimchi, yen tsai, zha cai
    - Central Asia: kumis (mare milk), kefir, shubat (camel milk)
    - South Asia: achar, appam, dosa, dhokla, dahi (yogurt), idli, kaanji, mixed pickle, ngari, hawaichaar, jaand (rice beer), sinki, tongba, paneer

- Africa: fermented millet porridge, garri, hibiscus seed, hot pepper sauce, injera, lamoun makbouss, laxoox, mauoloh, msir, mslalla, oilseed, ogi, ogili, ogiri, iru
- Americas: sourdough bread, cultured milk, chicha, elderberry wine, kombucha, pickling (pickled vegetables), sauerkraut, lupin seed, oilseed, chocolate, vanilla, tabasco, tibicos, pulque, mikyuk (fermented bowhead whale)
- Middle East: kushuk, lamoun makbouss, mekhalel, torshi, boza
- Europe: rakfisk, sauerkraut, pickled cucumber, surströmming, mead, elderberry wine, salami, sucuk, prosciutto, cultured milk products such as quark, kefir, filmjölk, crème fraîche, smetana, skyr, rakı, tupí.
- Oceania: poi, kaanga pirau (rotten corn), sago

Fermented Foods by Type

Bean-based

Cheonggukjang, doenjang, miso, natto, soy sauce, stinky tofu, tempeh, oncom, soybean paste, Beijing mung bean milk, kinama, iru

Dough-based

Proofing (baking technique)

Grain-based

Batter made from rice and lentil (Vigna mungo) prepared and fermented for baking idlis and dosas

Amazake, beer, bread, choujiu, gamju, injera, kvass, makgeolli, murri, ogi, rejuvelac, sake, sikhye, sourdough, sowans, rice wine, malt whisky, grain whisky, idli, dosa, vodka, boza

Vegetable-based

Kimchi, mixed pickle, sauerkraut, Indian pickle, gundruk, tursu

Fermenting cocoa beans

Fruit-based

Wine, vinegar, cider, perry, brandy, atchara, nata de coco, burong mangga, asinan, pickling, vişinată, chocolate, rakı

Honey-based

Mead, metheglin

Dairy-based

Cheeses in art: Still Life with Cheeses, Almonds and Pretzels, Clara Peeters, c. 1615

Some kinds of cheese also, kefir, kumis (mare milk), shubat (camel milk), cultured milk products such as quark, filmjölk, crème fraîche, smetana, skyr, and yogurt

Fish-based

Bagoong, faseekh, fish sauce, Garum, Hákarl, jeotgal, rakfisk, shrimp paste, surströmming, shidal

Meat-based

Chin som mok is a northern Thai speciality made with grilled, banana leaf-wrapped pork (both skin and meat) that has been fermented with glutinous rice

Chorizo, salami, sucuk, pepperoni, nem chua, som moo, saucisson

Tea-based

Pu-erh tea, Kombucha

Risks

Alaska has witnessed a steady increase of cases of botulism since 1985. It has more cases of botulism than any other state in the United States of America. This is caused by the traditional Eskimo practice of allowing animal products such as whole fish, fish heads, walrus, sea lion, and whale flippers, beaver tails, seal oil, and birds, to ferment for an extended period of time before being consumed. The risk is exacerbated when a plastic container is used for this purpose instead of the old-fashioned, traditional method, a grass-lined hole, as the *Clostridium botulinum* bacteria thrive in the anaerobic conditions created by the air-tight enclosure in plastic.

The World Health Organization has classified pickled foods as possibly carcinogenic, based on epidemiological studies. Other research found that fermented food contains a carcinogenic by-product, ethyl carbamate (urethane). "A 2009 review of the existing studies conducted across Asia concluded that regularly eating pickled vegetables roughly doubles a person's risk for esophageal squamous cell carcinoma."

Various Fermented Food

Yogurt

Yogurt, yoghurt, or yoghourt (from Turkish: *yoğurt*; other spellings listed below) is a food produced by bacterial fermentation of milk.

The bacteria used to make yogurt are known as "yogurt cultures". Fermentation of lactose by these bacteria produces lactic acid, which acts on milk protein to give yogurt its texture and characteristic tang. Cow's milk is commonly available worldwide, and, as such, is the milk most commonly used to make yogurt. Milk from water buffalo, goats, ewes, mares, camels, and yaks is also used to produce yogurt where available locally. Milk used may be homogenized or not (milk distributed in many parts of the world is homogenized); both types may be used, with substantially different results.

Yogurt is produced using a culture of *Lactobacillus delbrueckii* subsp. *bulgaricus* and *Streptococcus thermophilus* bacteria. In addition, other lactobacilli and bifidobacteria are also sometimes added during or after culturing yogurt. Some countries require yogurt to contain a certain amount of colony-forming units of bacteria; in China, for example, the requirement for the number of lactobacillus bacteria is at least 1×10^6 CFA per gram per milliliter. To produce yogurt, milk is first heated, usually to about 85 °C (185 °F), to denature the milk proteins so that they set together rather than form curds. After heating, the milk is allowed to cool to about 45 °C (113 °F). The bacterial culture is mixed in, and a temperature of 45 °C (113 °F) is maintained for four to seven hours to allow fermentation.

Etymology and Spelling

The word is derived from Turkish: *yoğurt*, and is usually related to the verb *yoğurmak*, "to knead", or "to be curdled or coagulated; to thicken". It may be related to *yoğun*, meaning thick or dense. The letter ğ was traditionally rendered as "gh" in transliterations of Turkish prior to 1928. In older Turkish, the letter denoted a voiced velar fricative, but this sound is elided between back vowels in modern Turkish, in which the word is pronounced.

In English, there are several variations of the spelling of the word, including *yogurt*, *yoghurt* and to a lesser extent *yoghourt*, *yogourt*, *yaghourt*, *yahourth*, *yoghurd*, *joghourt*, and *jogourt*. In the United Kingdom and Australia, *yogurt* and *yoghurt* are both current, *yogurt* being used by the Australian and British dairy councils, and *yoghourt* is an uncommon alternative. In the United States, Canada, and New Zealand, *yogurt* is the usual spelling and *yoghurt* a minor variant.

Historically, there have been cases of yogurt being spelled with a "J" instead of a "Y" (e.g. *jogurt* and *joghurt*) due to alternative transliteration methods. However, there has been a decline in these variations in English speaking countries; in certain European countries, on the other hand, it is still commonly spelled with a "J". Most people tend to spell in the manner shown on the packaging of the major brands in their country.

Whatever the spelling, the word is usually pronounced with a short in England and Wales, and with a long in Scotland, North America, Australia, New Zealand, Ireland and South Africa.

History

Analysis of the *L. delbrueckii* subsp. *bulgaricus* genome indicates that the bacterium may have originated on the surface of a plant. Milk may have become spontaneously and unintentionally exposed to it through contact with plants, or bacteria may have been transferred from the udder of domestic milk-producing animals.

The origins of yogurt are unknown, but it is thought to have been invented in Mesopotamia around 5000 BC.

In ancient Indian records, the combination of yogurt and honey is called "the food of the gods". Persian traditions hold that "Abraham owed his fecundity and longevity to the regular ingestion of yogurt".

The cuisine of ancient Greece included a dairy product known as oxygala which is believed to have been a form of yogurt. Galen (AD 129 – c. 200/c. 216) mentioned that oxygala was consumed with honey, similar to the way thickened Greek yogurt is eaten today.

Unstirred Turkish Süzme Yoğurt (strained yogurt), with a 10% fat content

The oldest writings mentioning yogurt are attributed to Pliny the Elder, who remarked that certain "barbarous nations" knew how "to thicken the milk into a substance with an agreeable acidity". The use of yogurt by medieval Turks is recorded in the books *Diwan Lughat al-Turk* by Mahmud Kashgari and *Kutadgu Bilig* by Yusuf Has Hajib written in the 11th century. Both texts mention the word "yogurt" in different sections and describe its use by nomadic Turks. The earliest yogurts were probably spontaneously fermented by wild bacteria in goat skin bags.

Some accounts suggest that Indian emperor Akbar's cooks would flavor yogurt with mustard seeds and cinnamon. Another early account of a European encounter with yogurt occurs in French clinical history: Francis I suffered from a severe diarrhea which no French doctor could cure. His ally Suleiman the Magnificent sent a doctor, who allegedly cured the patient with yogurt. Being grateful, the French king spread around the information about the food which had cured him.

Until the 1900s, yogurt was a staple in diets of people in the Russian Empire (and especially Central Asia and the Caucasus), Western Asia, South Eastern Europe/Balkans, Central Europe, and India. Stamen Grigorov (1878–1945), a Bulgarian student of medicine in Geneva, first examined the microflora of the Bulgarian yogurt. In 1905, he described it as consisting of a spherical and a rod-like lactic acid-producing bacteria. In 1907, the rod-like bacterium was called *Bacillus bulgaricus* (now *Lactobacillus delbrueckii subsp. bulgaricus*). The Russian Nobel laureate and biologist Ilya Ilyich Mechnikov (also known as Élie Metchnikoff), from the Institut Pasteur in Paris, was influenced by Grigorov's work and hypothesized that regular consumption of yogurt was responsible for the unusually long lifespans of Bulgarian peasants. Believing *Lactobacillus* to be essential for good health, Mechnikov worked to popularize yogurt as a foodstuff throughout Europe.

Isaac Carasso industrialized the production of yogurt. In 1919, Carasso, who was from Ottoman Salonika, started a small yogurt business in Barcelona, Spain, and named the business Danone ("little Daniel") after his son. The brand later expanded to the United States under an Americanized version of the name: Dannon.

Yogurt with added fruit jam was patented in 1933 by the Radlická Mlékárna dairy in Prague.

Yogurt was introduced to the United States in the first decade of the twentieth century, influenced by Élie Metchnikoff's *The Prolongation of Life; Optimistic Studies* (1908); it was available in tablet form for those with digestive intolerance and for home culturing. It was popularized by John Harvey Kellogg at the Battle Creek Sanitarium, where it was used both orally and in enemas, and later by Armenian immigrants Sarkis and Rose Colombosian, who started "Colombo and Sons Creamery" in Andover, Massachusetts in 1929. Colombo Yogurt was originally delivered around New England in a horse-drawn wagon inscribed with the Armenian word "madzoon" which was later changed to "yogurt", the Turkish name of the product, as Turkish was the lingua franca between immigrants of the various Near Eastern ethnicities who were the main consumers at that time. Yogurt's popularity in the United States was enhanced in the 1950s and 1960s, when it was presented as a health food. By the late 20th century, yogurt had become a common American food item and Colombo Yogurt was sold in 1993 to General Mills, which discontinued the brand in 2010.

Nutrition and Health

In a 100-gram amount providing 406 kilojoules (97 kcal) of dietary energy, yogurt (plain Greek yogurt from whole milk) is 81% water, 9% protein, 5% fat and 4% carbohydrates, including 4% sugars (table). As a proportion of the Daily Value (DV), a serving of yogurt is a rich source of vitamin B_{12} (31% DV) and riboflavin (23% DV), with moderate content of protein, phosphorus and selenium (14 to 19% DV; table).

Comparison of Whole Dairy Milk and Plain Yogurt from Whole Dairy Milk, One Cup (245 g) Each		
Property	**Milk**	**Yogurt**
kilo calories	146	149
Total Fat	7.9 g	8.5 g
Cholesterol	24.4 mg	11 mg
Sodium	98 mg	113 mg
Phosphorus	222 mg	233 mg
Potassium	349 mg	380 mg
Total Carbohydrates	12.8 g	12 g
Protein	7.9 g	9 g
Vitamin A	249 IU	243 IU
Vitamin C	0.0 mg	1.2 mg
Vitamin D	96.5 IU	~
Vitamin E	0.1 mg	0.1 mg
Vitamin K	0.5 µg	0.5 µg
Thiamine	0.1 mg	0.1 mg
Riboflavin	0.3 mg	0.3 mg
Niacin	0.3 mg	0.2 mg
Vitamin B6	0.1 mg	0.1 mg
Folate	12.2 µg	17.2 µg
Vitamin B12	1.1 µg	0.9 µg
Choline	34.9 mg	1.0 mg
Betaine	1.5 mg	~
Water	215 g	215 g
Ash	1.7 g	1.8 g

Tilde (~) represents missing or incomplete data. – The above shows that there is little difference between whole milk and yogurt made from whole milk with respect to the listed nutritional constituents. The differences may be explained as a result of testing the product after draining liquid whey from the yogurt thereby changing the percentage of that constituent in the final product.

Although yogurt is often associated with probiotics having positive effects on immune, cardiovascular or metabolic health, there is insufficient high-quality clinical evidence to conclude that consuming yogurt lowers risk of diseases or improves health.

Lactose-intolerant individuals may tolerate yogurt better than other dairy products due to the conversion of lactose to the sugars glucose and galactose, and the fermentation of lactose to lactic acid carried out by the bacteria present in the yogurt.

Varieties and Presentation

Tzatziki is a side dish made with yogurt, popular in Greek cuisine, and similar yet thicker than the Turkish Cacik and close to the traditional Bulgarian Milk salad.

Skyr is an Icelandic cultured dairy product, similar to strained yogurt. It has been a part of Icelandic cuisine for over a thousand years. It is traditionally served cold with milk and a topping of sugar.

Cacık, a Turkish cold appetizer made from yogurt

Da-hi is a yogurt of the Indian subcontinent, known for its characteristic taste and consistency. The word *da-hi* seems to be derived from the Sanskrit word *dadhi*, one of the five elixirs, or panchamrita, often used in Hindu ritual. *Dahi* also holds cultural symbolism in many homes in the *Mithila* region of Nepal and Bihar. Yogurt balances the palate across regional cuisines throughout India. In the hot and humid south, yogurt and foods made of yogurt are a staple in order to cool down – yogurt rice is always the last dish of the meal. Also, the vegetarian population of India derives some protein from yogurt (other than lentil and beans). Sweet yogurt (*meesti doi* or *meethi dahi*) is common in eastern parts of India, made by fermenting sweetened milk. While cow's milk is considered sacred and is currently the primary ingredient for yogurt, goat and buffalo milk were widely used in the past, and valued for the fat content. Butter and cream were made by churning the yogurt/milk.

Tarator is a cold soup made of yogurt, cucumber, dill, garlic and sunflower oil (walnuts are sometimes added) and is popular in Bulgaria.

In India and Pakistan, it is often used in cosmetics mixed with turmeric and honey. Sour yogurt, is also used as a hair conditioner by women in many parts of India and Pakistan. *Dahi* is also known as *Mosaru* (Kannada), *Thayir* (Tamil/Malayalam), *doi* (Assamese, Bengali), *dohi* (Odia), *perugu* (Telugu) and *Dhahi* or *Dhaunro*.

Raita is a condiment made with yogurt and popular in India, Pakistan and Bangladesh.

Dadiah sold in Bukittinggi Market

Raita is a yogurt-based South Asian/Indian condiment, used as a side dish. The yogurt is seasoned with coriander (cilantro), cumin, mint, cayenne pepper, and other herbs and spices. Vegetables such as cucumber and onions are mixed in, and the mixture is served chilled. Raita has a cooling effect on the palate which makes it a good foil for spicy Indian and Pakistani dishes. Raita is sometimes also referred to as *dahi.*

Dadiah or dadih is a traditional West Sumatran yogurt made from water buffalo milk, fermented in bamboo tubes.

Yogurt is popular in Nepal, where it is served as both an appetizer and dessert. Locally called *dahi,* it is a part of the Nepali culture, used in local festivals, marriage ceremonies, parties, religious occasions, family gatherings, and so on. The most famous type of Nepalese yogurt is called *juju dhau,* originating from the city of Bhaktapur.

In Tibet, yak milk (technically dri milk, as the word yak refers to the male animal) is made into yogurt (and butter and cheese) and consumed.

In Northern Iran, *Mâst Chekide* is a variety of kefir yogurt with a distinct sour taste. It is usually mixed with a pesto-like water and fresh herb purée called delal. Yogurt is a side dish to all Iranian meals. The most popular appetizers are spinach or eggplant borani, *Mâst-o-Khiâr* with cucumber, spring onions and herbs, and *Mâst-Musir* with wild shallots. In the summertime, yogurt and ice cubes are mixed together with cucumbers, raisins, salt, pepper and onions and topped with some croutons made of Persian traditional bread and served as a cold soup. Ashe-Mâst is a warm yogurt soup with fresh herbs, spinach and lentils. Even the leftover water extracted when straining yogurt is cooked to make a sour cream sauce called kashk, which is usually used as a topping on soups and stews.

Matsoni is a Georgian yogurt popular in the Caucasus and Russia. It is used in a wide variety of Georgian dishes and is believed to have contributed to the high life expectancy and longevity in the country. Dannon used this theory in their 1978 TV advertisement called *In Soviet Georgia* where shots of elderly Georgian farmers were interspersed with an off-camera announcer intoning, "In Soviet Georgia, where they eat a lot of yogurt, a lot of people live past 100." Matsoni is also popular in Japan under the name Caspian Sea Yogurt.

Tarator and Cacık are popular cold soups made from yogurt, popular during summertime in Albania, Azerbaijan (known as Dogramac), Bulgaria, Macedonia, Serbia and Turkey. They are made with ayran, cucumbers, dill, salt, olive oil, and optionally garlic and ground walnuts. Tzatziki in Greece and milk salad in Bulgaria are thick yogurt-based salads similar to tarator.

Khyar w Laban (cucumber and yogurt salad) is a popular dish in Lebanon and Syria. Also, a wide variety of local Lebanese and Syrian dishes are cooked with yogurt like "Kibbi bi Laban", etc.

Rahmjoghurt, a creamy yogurt with much higher fat content (10%) than many yogurts offered in English-speaking countries (*Rahm* is German for "cream"), is available in Germany and other countries.

Dovga, a yogurt soup cooked with a variety of herbs and rice is popular in Azerbaijan, often served warm in winter or refreshingly cold in summer.

Yogurt made with unhomogenized milk is sometimes called cream-top yogurt; a layer of cream rises to the top.

Jameed is yogurt which is salted and dried to preserve it. It is popular in Jordan.

Zabadi is the type of yogurt made in Egypt, usually from the milk of the Egyptian water buffalo. It is particularly associated with Ramadan fasting, as it is thought to prevent thirst during all-day fasting.

Sweetened and Flavored Yogurt

To offset its natural sourness, yogurt is also sold sweetened, flavored or in containers with fruit or fruit jam on the bottom. The two styles of yogurt commonly found in the grocery store are set type yogurt and Swiss style yogurt. Set type yogurt is when the yogurt is packaged with the fruit on the bottom of the cup and the yogurt on top. Swiss style yogurt is when the fruit is blended into the yogurt prior to packaging.

Lassi and Moru are common beverages in India. Lassi is milk that is sweetened with sugar commonly, less commonly honey and often combined with fruit pulp to create flavored lassi. Mango lassi is a western favorite, as is coconut lassi. Consistency can vary widely, with urban and commercial lassis being of uniform texture through being processed, whereas rural and rustic lassi has curds in it, and sometimes has malai (cream) added or removed. Moru is a popular South Indian summer drink, meant to keep drinkers hydrated through the hot and humid summers of the South. It is prepared by considerably thinning down yogurt with water, adding salt (for electrolyte balance) and spices, usually green chili peppers, asafoetida, curry leaves and mustard.

Large amounts of sugar – or other sweeteners for low-energy yogurts – are often used in commercial yogurt. Some yogurts contain added starch, pectin (found naturally in fruit), and/or gelatin to create thickness and creaminess artificially at lower cost. Gelatin is a meat or fish product, therefore vegetarians should avoid products containing it. This type of yogurt is also marketed under the name Swiss-style, although it is unrelated to the way yogurt is eaten in Switzerland. Some yogurts, often called "cream line", are made with whole milk which has not been homogenized so the cream rises to the top.

In the UK, Ireland, France and United States, sweetened, flavored yogurt is the most popular type, typically sold in single-serving plastic cups. Common flavors include vanilla, honey, and toffee, and fruit such as strawberry, cherry, blueberry, blackberry, raspberry, mango and peach. In the early twenty-first century yogurt flavors inspired by desserts, such as chocolate or cheesecake, have been available.

There is concern about the health effects of sweetened yogurt. The United Kingdom and the United States recommend different maximum amounts of daily sugar intake but in both nations many sweetened yogurts have too much.

A 150g (5oz) serving of some 0% fat yogurts can contain as much as 20g (0.7oz) of sugar – the equivalent of five teaspoons, says Action on Sugar – which is about 40% of a woman's daily recommended intake of added sugar (50g or 1.7oz) and about 30% of that for men (70g or 2.5oz).

The American Heart Association recommends that men eat no more than 36 grams of sugar per day, and women no more than 20. Many of the top-selling yogurts have even more than the 19 grams of sugar that is contained in a Twinkie.

Consumers wanting sweetened yogurt are advised to choose yogurt sweetened with sugar substitute and check the contents list to avoid corn syrup, high fructose corn syrup, honey, or sugar.

Strained Yogurt

Strained yogurt is yogurt which has been strained through a filter, traditionally made of muslin and more recently of paper or cloth. This removes the whey, giving a much thicker consistency and a distinctive slightly tangy taste. Strained yogurt is becoming more popular with those who make yogurt at home, especially if using skimmed milk which results in a thinner consistency.

Yogurt that has been strained to filter or remove the whey is known as Labneh in Middle Eastern countries. It has a consistency between that of yogurt and cheese. It is popular for sandwiches in Arab countries. Olive oil, cucumber slices, olives, and various green herbs may be added. It can be thickened further and rolled into balls, preserved in olive oil, and fermented for a few more weeks. It is sometimes used with onions, meat, and nuts as a stuffing for a variety of pies or kibbeh balls.

Some types of strained yogurts are boiled in open vats first, so that the liquid content is reduced. The popular East Indian dessert, a variation of traditional dahi called mishti dahi, offers a thicker, more custard-like consistency, and is usually sweeter than western yogurts.

Use coffee filter to strain yogurt in a home refrigerator.

Strained yogurt is also enjoyed in Greece and is the main component of *tzatziki* (from Turkish "cacık"), a well-known accompaniment to gyros and souvlaki pita sandwiches: it is a yogurt sauce or dip made with the addition of grated cucumber, olive oil, salt and, optionally, mashed garlic.

Srikhand, a popular dessert in India, is made from strained yogurt, saffron, cardamom, nutmeg and sugar and sometimes fruits such as mango or pineapple.

In North America and Britain, strained yogurt is commonly called "Greek yogurt". Strained yogurt is sometimes marketed in North America as "Greek yogurt" and in Britain as "Greek-style yoghurt". In Britain the name "Greek" may only be applied to yogurt made in Greece.

Beverages

Dugh ("dawghe" in Neo-Aramaic), ayran or dhallë is a yogurt-based, salty drink popular in Iran, Albania, Bulgaria, Turkey, Azerbaijan, Afghanistan, Pakistan, Bangladesh, Macedonia, Uzbekistan, Kazakhstan and Kyrgyzstan. It is made by mixing yogurt with water and (sometimes) salt. The same drink is known as *doogh* in Iran, in Armenia; *laban ayran* in Syria and Lebanon; *shenina* in Iraq and Jordan; *laban arbil* in Iraq; *majjiga* (Telugu), *majjige* (Kannada), and *moru* (Tamil and Malayalam) in South India; namkeen *lassi* in Punjab and all over Pakistan. A similar drink, doogh, is popular in the Middle East between Lebanon, Iran, and Iraq; it differs from ayran by the addition of herbs, usually mint, and is sometimes carbonated, commonly with carbonated water.

PCC Dairy Yogurt Milk, with live cultures, made from water buffalo's cream milk Philippine Carabao Center.

Borhani (or Burhani) is a spicy yogurt drink popular in Bangladesh and parts of Bengal. It is usually served with kacchi biryani at weddings and special feasts. Key ingredients are yogurt blended with mint leaves (mentha), mustard seeds and black rock salt (Kala Namak). Ground roasted cumin, ground white pepper, green chili pepper paste and sugar are often added.

Ayran. One of the popular beverages in Turkish cuisine.

Lassi is a yogurt-based beverage originally from the Indian subcontinent that is usually slightly salty or sweet. Lassi is a staple in Punjab. In some parts of the subcontinent, the sweet version may be commercially flavored with rosewater, mango or other fruit juice to create a very different drink. Salty lassi is usually flavored with ground, roasted cumin and red chilies; this salty variation may also use buttermilk, and in India is interchangeably called *ghol* (Bengal), *mattha* (North India), "majjige" (Karnataka), *majjiga* (Telangana & Andhra Pradesh), *moru* (Tamil Nadu and Kerala), *Dahi paani Chalha* (Odisha), *tak* (Maharashtra), or *chaas* (Gujarat). Lassi is very widely drunk in India, Pakistan, and Bangladesh. Mango Lassi is a popular drink at Indian restaurants in US.

In Bosnia and Herzegovina, Croatia, Macedonia, Montenegro, Serbia, and Slovenia, an unsweetened and unsalted yogurt drink usually called simply *jogurt* is a popular accompaniment to *burek* and other baked goods.

Sweetened yogurt drinks are the usual form in Europe (including the UK) and the US, containing fruit and added sweeteners. These are typically called "drinkable yogurt".

Also available are "yogurt smoothies", which contain a higher proportion of fruit and are more like smoothies. In Ecuador, yogurt smoothies flavored with native fruit are served with pan de yuca as a common type of fast food.

Also in Turkey, yogurt soup or *Yayla Çorbası* is a popular way of consuming yogurt. The soup is a mix of yogurt, rice, flour and dried mint.

Plant-milk Yogurt

Plant-milk yogurt

A variety of plant-milk yogurts appeared in the 2000s, using soy milk, rice milk, and nut milks such as almond milk and coconut milk. So far the most widely sold variety of plant milk yogurts is soy yogurt. These yogurts are suitable for vegans, people with intolerance to dairy milk, and those who prefer plant milks.

Making Yogurt at Home

Commercially available yogurt maker

Yogurt is made by heating milk to a temperature that denaturates its proteins (scalding), essential for making yogurt, cooling it to a temperature that will not kill the live microorganisms that turn the milk into yogurt, inoculating certain bacteria (starter culture), usually *Streptococcus thermophilus* and *Lactobacillus bulgaricus*, into the milk, and finally keeping it warm for several hours. The milk may be held at 85 °C (185 °F) for a few minutes, or boiled (giving a somewhat different result). It must be cooled to 50 °C (122 °F) or somewhat less, typically 40–46 °C (104–115 °F). Starter culture must then be mixed in well, and the mixture must be kept undisturbed and warm for several hours, ranging from 5 to 12, with longer fermentation producing a more acid yogurt. The starter culture may be a small amount of live yogurt; dried starter culture is available commercially.

Home yogurt maker

Milk with a higher concentration of solids than normal milk may be used; the higher solids content produces a firmer yogurt. Solids can be increased by adding dried milk.

The yogurt-making process provides two significant barriers to pathogen growth, heat and acidity (low pH). Both are necessary to ensure a safe product. Acidity alone has been questioned by recent outbreaks of food poisoning by *E. coli O157:H7* that is acid-tolerant. *E. coli O157:H7* is easily destroyed by pasteurization (heating); the initial heating of the milk kills pathogens as well as denaturing proteins. The microorganisms that turn milk into yogurt can tolerate higher temperatures than most pathogens, so that a suitable temperature not only encourages the formation of yogurt, but inhibits pathogenic microorganisms.

Once the yogurt has formed it can, if desired, be strained to reduce the whey content and thicken it.

Vinegar

Vinegar is a liquid consisting of about 5–20% acetic acid (CH_3COOH), water, and other trace chemicals, which may include flavorings. The acetic acid is produced by the fermentation of ethanol by acetic acid bacteria. Vinegar is now mainly used as a cooking ingredient, or in pickling. As the most easily manufactured mild acid, it has historically had a great variety of industrial, medical, and domestic uses, some of which (such as its use as a general household cleaner) are still commonly practiced today.

A variety of flavored vinegars on sale in France (bottom rows)

Commercial vinegar is produced either by fast or slow fermentation processes. In general, slow methods are used with traditional vinegars, and fermentation proceeds slowly over the course of months or a year. The longer fermentation period allows for the accumulation of a nontoxic slime composed of acetic acid bacteria. Fast methods add mother of vinegar (bacterial culture) to the source liquid before adding air to oxygenate and promote the fastest fermentation. In fast production processes, vinegar may be produced in 20 hours to three days.

History

Vinegar has been made and used by people for thousands of years. Traces of it have been found in Egyptian urns from around 3000 BC.

Varieties

Apple Cider

Apple cider vinegar is made from cider or apple must, and has a brownish-gold color. It is often sold unfiltered and unpasteurized with the mother of vinegar present, as a natural product. It is often diluted with fruit juice or water or sweetened (usually with honey) for consumption as a health beverage.

Balsamic

Balsamic vinegar is an aromatic aged vinegar produced in the Modena and Reggio Emilia provinces of Italy. The original product—Traditional Balsamic Vinegar—is made from the concentrated juice, or must, of white Trebbiano grapes. It is very dark brown, rich, sweet, and complex, with the finest grades being aged in successive casks made variously of oak, mulberry, chestnut, cherry, juniper, and ash wood. Originally a costly product available to only the Italian upper classes, traditional balsamic vinegar is marked "tradizionale" or "DOC" to denote its Protected Designation of Origin status, and is aged for 12 to 25 years. A cheaper non-DOC commercial form described as "aceto balsamico di Modena" (balsamic vinegar of Modena) became widely known and available around the world in the late 20th century, typically made with concentrated grape juice mixed with a strong vinegar, then coloured and slightly sweetened with caramel and sugar.

Regardless of how it is produced, balsamic vinegar must be made from a grape product. It contains no balsam fruit. A high acidity level is somewhat hidden by the sweetness of the other ingredients, making it very mellow.

Beer

Vinegar made from beer is produced in the United Kingdom, Germany, Austria, and the Netherlands. Although its flavor depends on the particular type of beer from which it is made, it is often described as having a malty taste. That produced in Bavaria is a light golden color with a very sharp and not-overly-complex flavor.

Cane

Cane vinegar, made from sugarcane juice, is most popular in the Philippines, in particular, the Ilocos Region of the northern Philippines (where it is called *sukang iloko*), although it also is produced in France and the United States. It ranges from dark yellow to golden brown in color, and has a mellow flavor, similar in some respects to rice vinegar, though with a somewhat "fresher" taste. Because it contains no residual sugar, it is no sweeter than any other vinegar. In the Philippines, it often is labeled as *sukang maasim* (Tagalog for "sour vinegar").

Cane vinegars from Ilocos are made in two different ways. One way is to simply place sugar cane juice in large jars and it will become sour by the direct action of bacteria on the sugar. The other way is through fermentation to produce a local wine known as *basi*. Low-quality *basi* is then allowed to undergo acetic acid fermentation that converts alcohol into acetic acid. Contaminated *basi* also becomes vinegar.

A white variation has become quite popular in Brazil in recent years, where it is the cheapest type of vinegar sold. It is now common for other types of vinegar (made from wine, rice and apple cider) to be sold mixed with cane vinegar to lower the cost.

Sugarcane sirka is made from sugarcane juice in Punjab, India. During summer people put cane juice in earthenware pots with iron nails. The fermentation takes place due to the action of wild yeast. The cane juice is converted to vinegar having a blackish color. The sirka is used to preserve pickles and for flavoring curries.

Coconut

Coconut vinegar, made from fermented coconut water or sap, is used extensively in Southeast Asian cuisine (notably the Philippines), as well as in some cuisines of India and Sri Lanka, especially Goan cuisine. A cloudy white liquid, it has a particularly sharp, acidic taste with a slightly yeasty note.

Date

Vinegar made from dates is a traditional product of the Middle East .

Distilled

The term "distilled vinegar" is something of a misnomer when used in the US and North America, because it is not produced by distillation but by fermentation of distilled alcohol. The fermentate is diluted to produce a colorless solution of 5% to 8% acetic acid in water, with a pH of about 2.6. This is variously known as distilled spirit, "virgin" vinegar, or white vinegar, and is used in cooking, baking, meat preservation, and pickling, as well as for medicinal, laboratory, and cleaning purposes. The most common starting material in some regions, because of its low cost, is malt, or in the United States, corn (maize). It is sometimes derived from petroleum. Distilled vinegar in the UK is produced by the distillation of malt to give a clear vinegar which maintains some of the malt flavour. Distilled vinegar is used predominantly for cooking, although in Scotland it is used as an alternative to brown or light malt vinegar. White distilled vinegar can also be used for cleaning.

East Asian Black

Chinese black vinegar

Chinese black vinegar is an aged product made from rice, wheat, millet, sorghum, or a combination thereof. It has an inky black color and a complex, malty flavor. There is no fixed recipe, so some Chinese black vinegars may contain added sugar, spices, or caramel color. The most popular variety, Zhenjiang vinegar, originates in the city of Zhenjiang in Jiangsu Province, eastern China. Shanxi mature vinegar is another popular type of Chinese vinegar that is made exclusively from sorghum and other grains Nowadays in Shanxi province, there are still some traditional vinegar workshops producing handmade vinegar which aged for at least five years with a high acidity. Only the vinegar made in Taiyuan and some counties in Jinzhong and aged for at least three years is considered authentic Shanxi mature vinegar according to the latest national standard.

A somewhat lighter form of black vinegar, made from rice, is produced in Japan, where it is called *kurozu*.

Fruit

Persimmon vinegar produced in South Korea

Fruit vinegars are made from fruit wines, usually without any additional flavoring. Common flavors of fruit vinegar include apple, blackcurrant, raspberry, quince, and tomato. Typically, the flavors of the original fruits remain in the final product.

Most fruit vinegars are produced in Europe, where there is a growing market for high-price vinegars made solely from specific fruits (as opposed to non-fruit vinegars that are infused with fruits or fruit flavors). Several varieties, however, also are produced in Asia. Persimmon vinegar, called *gam sikcho*, is popular in South Korea. Jujube vinegar, called *zaocu* or *hongzaocu*, and wolfberry vinegar are produced in China.

Honey

Vinegar made from honey is rare, although commercially available honey vinegars are produced in Italy, Portugal, France, Romania, and Spain.

Job's Tears

In Japan, an aged vinegar also is made from Job's tears, a tall, grain-bearing, tropical plant. The vinegar is similar in flavor to rice vinegar.

Kiwifruit

A byproduct of commercial kiwifruit growing is a large amount of waste in the form of misshapen or otherwise-rejected fruit (which may constitute up to 30 percent of the crop) and kiwifruit pomace, the presscake residue left after kiwifruit juice manufacture. One of the uses for this waste is the production of kiwifruit vinegar, produced commercially in New Zealand since at least the early 1990s, and in China in 2008.

Kombucha

Kombucha vinegar is made from kombucha, a symbiotic culture of yeast and bacteria. The bacteria produce a complex array of nutrients and populate the vinegar with bacteria that some claim promote a healthy digestive tract, although no scientific studies have confirmed this. Kombucha vinegar primarily is used to make a vinaigrette, and is flavored by adding strawberries, blackberries, mint, or blueberries at the beginning of fermentation.

Malt

Malt vinegar, also called alegar, is made by malting barley, causing the starch in the grain to turn to maltose. Then an ale is brewed from the maltose and allowed to turn into vinegar, which is then aged. It is typically light-brown in color. In the United Kingdom and Canada, malt vinegar (along with salt) is a traditional seasoning for fish and chips, but some commercial fish and chip shops use non-brewed condiment.

Palm

Palm vinegar, made from the fermented sap from flower clusters of the nipa palm (also called attap palm), is used most often in the Philippines, where it is produced, and where it is called *sukang paombong*. It has a citrusy flavor note to it and imparts a distinctly musky aroma. Its pH is between five and six.

Pomegranate

Pomegranate vinegar (Hebrew: חומץ רימונים) is used widely in Israel as a dress for salad but also in meat stew and in dips.

Raisin

Vinegar made from raisins (Arabic: خل عنب "grape vinegar") is used in cuisines of the

Middle East, and is produced there. It is cloudy and medium brown in color, with a mild flavor.

Raisin vinegar

Rice

Rice vinegar is most popular in the cuisines of East and Southeast Asia. It is available in "white" (light yellow), red, and black varieties. The Japanese prefer a light rice vinegar for the preparation of sushi rice and salad dressings. Red rice vinegar traditionally is colored with red yeast rice. Black rice vinegar (made with black glutinous rice) is most popular in China, and it is also widely used in other East Asian countries.

White rice vinegar has a mild acidity with a somewhat "flat" and uncomplex flavor. Some varieties of rice vinegar are sweetened or otherwise seasoned with spices or other added flavorings.

Sherry

Sherry vinegar

Sherry vinegar is linked to the production of sherrywines of Jerez. Dark-mahogany in color, it is made exclusively from the acetic fermentation of wines. It is concentrated and has generous aromas, including a note of wood, ideal for vinaigrettes and flavoring various foods.

Spirit

The term 'spirit vinegar' is sometimes reserved for the stronger variety (5% to 21% acetic acid) made from sugar cane or from chemically produced acetic acid. To be called "Spirit Vinegar", the product must come from an agricultural source and must be made by "double fermentation". The first fermentation is sugar to alcohol and the second alcohol to acetic acid. Product made from chemically produced acetic acid cannot be called "vinegar". In the UK the term allowed is "Non-brewed condiment".

Wine

Wine vinegar is made from red or white wine, and is the most commonly used vinegar in Southern and Central Europe, Cyprus and Israel. As with wine, there is a considerable range in quality. Better-quality wine vinegars are matured in wood for up to two years, and exhibit a complex, mellow flavor. Wine vinegar tends to have a lower acidity than white or cider vinegars. More expensive wine vinegars are made from individual varieties of wine, such as champagne, sherry, or pinot gris.

Uses

Culinary

Vinegar is commonly used in food preparation, in particular in pickling processes, vinaigrettes, and other salad dressings. It is an ingredient in sauces such as mustard, ketchup, and mayonnaise. Vinegar is sometimes used while making chutneys. It is often used as a condiment. Marinades often contain vinegar. In terms of its shelf life, vinegar's acidic nature allows it to last indefinitely without the use of refrigeration.

- Condiment for beetroot – cold, cooked beetroot is commonly eaten with vinegar and other ingredients
- Condiment for fish and chips (UK: chips; US: French fries) – in Britain, Ireland, Canada, and Australia, salt and malt vinegar is sprinkled on chips. In Canada, white vinegar is also often used.
- Flavoring for potato chips (UK: crisps) – many American, Canadian, British, and Australian manufacturers of packaged potato chips include a variety flavored with vinegar and salt.

- Vinegar pie – a North American variant on the dessert called chess pie. It is flavored with a small amount of cider vinegar and some versions also contain raisins, spices and sour cream.
- Pickling – any vinegar can be used to pickle foods.
- Cider vinegar and sauces – cider vinegar usually is not suitable for use in delicate sauces.
- Apple cider vinegar – Usually placed on the table in small bowls or cups so that people can dip their crab meat into it. Also mixed with water and used to steam crabs.
- Substitute for fresh lemon juice – cider vinegar can usually be substituted for fresh lemon juice in recipes and obtain a pleasing effect although it lacks the vitamin C.
- Saucing roast lamb – pouring cider vinegar over the meat when roasting lamb, especially when combined with honey or when sliced onions have been added to the roasting pan, produces a sauce.
- Sweetened vinegar is used in the dish of pork knuckles and ginger stew, which is made among Chinese people of Cantonese backgrounds to celebrate the arrival of a new child.
- Sushi rice – Japanese use rice vinegar as an essential ingredient for sushi rice.
- Commonly put into water-pepper sauce, for general palate preference.
- Red vinegar – Sometimes used in Chinese soups.
- Flavoring – used in the Southern U.S. to flavor collard greens, green beans, black-eyed peas, or cabbage to taste.
- Commonly put into mint sauce, for general palate preference.
- Vinegar – especially the coconut, cane, or palm variety – is one of the principal ingredients of Philippine cuisine.
- White vinegar can be used as flavoring in ham and beans.
- It is used in the making of escabeche fish.

Beverage

Several beverages are made using vinegar, for instance Posca. The ancient Greek oxymel is a drink made from vinegar and honey, and sekanjabin is a traditional Persian drink similar to oxymel. Other preparations range from simply mixing sugar water or honey water with small amounts of fruity vinegar, to making syrup by laying fruit or mint in vinegar essence for several days, then sieving off solid parts, and adding considerable amounts of sugar. Some prefer to also boil the result as a final step. These recipes have lost much of their popularity with the rise of carbonated beverages, such as soft drinks.

Folk Medicine and Research

Many traditional remedies and treatments have been ascribed to vinegar over millennia and in many different cultures, although no medical uses are verified in controlled clinical trials. Some folk medicine uses have side effects that represent health risks.

Diet and Diabetes

Small amounts of vinegar (approximately 25 g of domestic vinegar) added to food, or taken along with a meal, were proposed in preliminary research to reduce the glycemic index of carbohydrate food for people with and without diabetes.

Some preliminary research indicates that taking vinegar with food increases satiety and reduces the amount of food consumed.

Antimicrobial

The growth of several common foodborne pathogens sensitive to acidity is inhibited by common vinegar (5% acetic acid).

Among these are:

- Clostridium botulinum (which can survive anaerobic conditions and high temperatures and is thus not always affected by sterilization)
- Salmonella
- Listeria
- Staphylococcus
- Escherichia coli O157:H7 (which can tolerate an acidic environment to some extent)

The active ingredient in vinegar, acetic acid, can effectively kill mycobacteria, as tested against drug-resistant tuberculosis bacteria as well as other mycobacteria.

Polyphenols

The phenolic composition analysis of vinegar shows the presence of numerous flavonoids, phenolic acids and aldehydes.

Other Uses

Applying vinegar to common jellyfish stings deactivates the nematocysts, although not as effectively as hot water. This does not apply to the Portuguese man o' war, which, although generally considered to be a jellyfish, is not; vinegar applied to Portuguese man o' war stings can cause their nematocysts to discharge venom, making the pain worse.

Vinegar is not effective against lice. Combined with 60% salicylic acid, it is significantly more effective than placebo for the treatment of warts.

Potential Hazards

Like other acids, the acetic acid in vinegar attacks the enamel of the teeth and will cause decay and sensitivity in the teeth. As with other acids, some organizations recommend minimizing consumption, minimizing time in the mouth, not swirling it in the mouth, and counteracting the effects by using a baking soda mouth rinse.

Esophageal injury by apple cider vinegar tablets has been reported, and, because vinegar products sold for medicinal purposes are neither regulated nor standardized, they vary widely in content, pH, and other respects. Long-term heavy vinegar ingestion has one recorded case of possibly causing hypokalemia, hyperreninemia, and osteoporosis.

Cleaning

White vinegar is often used as a household cleaning agent. Because it is acidic, it can dissolve mineral deposits from glass, coffee makers, and other smooth surfaces. For most uses, dilution with water is recommended for safety and to avoid damaging the surfaces being cleaned.

Vinegar is an excellent solvent for cleaning epoxy resin and hardener, even after the epoxy has begun to harden. Malt vinegar sprinkled onto crumpled newspaper is a traditional, and still-popular, method of cleaning grease-smeared windows and mirrors in the United Kingdom. Vinegar can be used for polishing brass or bronze. Vinegar is widely known as an effective cleaner of stainless steel and glass.

Vinegar has been reputed to have strong antibacterial properties. One test by Good Housekeeping's microbiologist found that 5% vinegar is 90% effective against mold and 99.9% effective against bacteria, though another study showed that vinegar is less effective than Clorox and Lysol against poliovirus. In modern times experts have advised against using vinegar as a household disinfectant against human pathogens, as it is less effective than chemical disinfectants.

Vinegar is ideal for washing produce because it breaks down the wax coating and kills bacteria and mold. The editors of Cook's Illustrated found vinegar to be the most effective and safest way to wash fruits and vegetables, beating antibacterial soap, water and just a scrub brush in removing bacteria.

Vinegar has been marketed as an environmentally-friendly solution for many household cleaning problems. For example, vinegar has been cited recently as an eco-friendly urine cleaner for pets.

Vinegar is effective in removing clogs from drains, polishing silver, copper and brass as well as ungluing sticker-type price tags. Vinegar is one of the best ways to restore color to upholstery like curtains and carpet.

Vinegar also can help remove wallpaper. If the paper is coated with a mixture of vinegar and boiling water, it breaks down the glue for easy removal.

Agricultural and Horticultural

20% acetic acid vinegar can be used as a herbicide. Acetic acid is not absorbed into root systems; the vinegar will kill top growth, but perennial plants may reshoot.

Miscellaneous

Most commercial vinegar solutions available to consumers for household use do not exceed 5%. Solutions above 10% require careful handling, as they are corrosive and damaging to the skin.

When a bottle of vinegar is opened, mother of vinegar may develop. It is considered harmless and can be removed by filtering.

Vinegar eels (*Turbatrix aceti*), a form of nematode, may occur in some forms of vinegar unless the vinegar is kept covered. These feed on the mother of vinegar and can occur in naturally fermenting vinegar.

Some countries prohibit the selling of vinegar over a certain percentage acidity. As an example, the government of Canada limits the acetic acid of vinegars to between 4.1% and 12.3%.

According to legend, in France during the Black Plague, four thieves were able to rob houses of plague victims without being infected themselves. When finally caught, the judge offered to grant the men their freedom, on the condition that they revealed how they managed to stay healthy. They claimed that a medicine woman sold them a potion made of garlic soaked in soured red wine (vinegar). Variants of the recipe, called Four Thieves Vinegar, have been passed down for hundreds of years and are a staple of New Orleans hoodoo practices.

A solution of vinegar can be used for water slide decal application as used on scale models and musical instruments, among other things. One part white distilled vinegar (5%

acidity) diluted with two parts of distilled or filtered water creates a suitable solution for the application of water-slide decals to hard surfaces. The solution is very similar to the commercial products, often described as "decal softener", sold by hobby shops. The slight acidity of the solution softens the decal and enhances its flexibility, permitting the decal to cling to contours more efficiently.

When baking soda and vinegar are combined, the bicarbonate ion of the baking soda reacts to form carbonic acid, which decomposes into carbon dioxide and water.

Olive

The olive, known by the botanical name *Olea europaea,* meaning "european olive", is a species of small tree in the family Oleaceae, found in the Mediterranean Basin from Portugal to the Levant, the Arabian Peninsula, and southern Asia as far east as China, as well as the Canary Islands, Mauritius and Réunion. The species is cultivated in many places and considered naturalized in all the countries of the Mediterranean coast, as well as in Argentina, Saudi Arabia, Java, Norfolk Island, California and Bermuda.

Olea europeana sylvestris is a subspecies that corresponds to a smaller tree bearing noticeably smaller fruits.

The olive's fruit, also called the olive, is of major agricultural importance in the Mediterranean region as the source of olive oil; it is one of the three core ingredients in Mediterranean cuisine. The tree and its fruit give their name to the plant family, which also includes species such as lilacs, jasmine, *Forsythia* and the true ash trees (*Fraxinus*). The word derives from Latin *ŏlīva* ("olive fruit", "olive tree"; "olive oil" is *ŏlĕum*) a borrowing from the Greek "olive fruit", "olive tree") and λαιον (*élaion*, "olive oil") in the archaic form. The oldest attested forms of the Greek words are the Mycenaean, *e-ra-wa*, and *e-ra-wo*, *e-rai-wo*, written in the Linear B syllabic script. The word "oil" in multiple languages ultimately derives from the name of this tree and its fruit.

Description

The olive tree, *Olea europaea*, is an evergreen tree or shrub native to the Mediterranean, Asia and Africa. It is short and squat, and rarely exceeds 8–15 m (26–49 ft) in height. The *Pisciottana*, a unique variety comprising 40,000 trees found only in the area around Pisciotta in the Campania region of southern Italy often exceeds this, with correspondingly large trunk diameters. The silvery green leaves are oblong, measuring 4–10 cm (1.6–3.9 in) long and 1–3 cm (0.39–1.18 in) wide. The trunk is typically gnarled and twisted.

The small white, feathery flowers, with ten-cleft calyx and corolla, two stamens and bifid stigma, are borne generally on the previous year's wood, in racemes springing from the axils of the leaves.

19th-century illustrations

The fruit is a small drupe 1–2.5 cm (0.39–0.98 in) long, thinner-fleshed and smaller in wild plants than in orchard cultivars. Olives are harvested in the green to purple stage. Canned black olives have often been artificially blackened and may contain the chemical ferrous gluconate to improve the appearance. *Olea euro-paea* contains a seed commonly referred to in American English as a pit or a rock, and in British English as a stone.

Taxonomy

There are six natural subspecies of *Olea europaea* distributed over a wide range:

- *Olea europaea* subsp. *europaea* (Mediterranean Basin)
- *Olea europaea* subsp. *cuspidata* (from South Africa throughout East Africa, Arabia to South West China)
- *Olea europaea* subsp. *guanchica* (Canaries)
- *Olea europaea* subsp. *cerasiformis* (Madeira)
- *Olea europaea* subsp. *maroccana* (Morocco)
- *Olea europaea* subsp. *laperrinei* (Algeria, Sudan, Niger)

The subspecies *maroccana* and *cerasiformis* are respectively hexaploid and tetraploid.

Wild growing forms of the olive are sometimes treated as the species *Olea oleaster*.

Cultivars

There are hundreds of cultivars of the olive tree (*Olea europaea*). An olive's cultivar has a significant impact on its colour, size, shape, and growth characteristics, as well as the qualities of olive oil. Olive cultivars may be used primarily for oil, eating, or both. Olives cultivated for consumption are generally referred to as table olives.

Since many olive cultivars are self-sterile or nearly so, they are generally planted in pairs with a single primary cultivar and a secondary cultivar selected for its ability to fertilize the primary one. In recent times, efforts have been directed at producing hybrid cultivars with qualities such as resistance to disease, quick growth and larger or more consistent crops.

History

Prehistory

It seems certain that the olive tree as we know it today had its origin approximately 6,000 to 7,000 years ago in the region corresponding to ancient Persia and Mesopotamia. The olive plant later spread from these areas to the Levant.

The edible olive seems to have coexisted with humans for about 5,000 to 6,000 years, going back to the early Bronze Age (3150 to 1200 BC). Its origin can be traced to the Levant based on written tablets, olive pits, and wood fragments found in ancient tombs. At least one cookbook writer writes that the most ancient evidence of olive cultivation is found in Syria, Israel, and Crete.

The immediate ancestry of the cultivated olive is unknown. It is assumedthat *Olea europaea* may have arisen from *O. chrysophylla* in northern tropical Africa and that it was introduced into the countries of the Mediterranean Basin via Egypt and then Crete or the Levant, Tunisia and Asia Minor. Fossil Olea pollen has been found in Macedonia, and other places around the Mediterranean, indicating that this genus is an original element of the Mediterranean flora. Fossilized leaves of Olea were found in the palaeosols of the volcanic Greek island of Santorini (Thera) and were dated about 37,000 BP. Imprints of larvae of olive whitefly *Aleurolobus (Aleurodes) olivinus* were found on the leaves. The same insect is commonly found today on olive leaves, showing that the plant-animal co-evolutionary relations have not changed since that time. Other leaves found on the same island are dated back to 60,000 BP, making them the oldest known olives from the Mediterranean.

As far back as 3000 BC, olives were grown commercially in Crete; they may have been the source of the wealth of the Minoan civilization.

Outside the Mediterranean

Olives are not native to the Americas. Spanish colonists brought the olive to the New World where its cultivation prospered in present-day Peru and Chile. The first seedlings from Spain were planted in Lima by Antonio de Rivera in 1560. Olive tree cultivation quickly spread along the valleys of South America's dry Pacific coast where the climate was similar to the Mediterranean. Spanish missionaries established the tree in the 18th century in California. It was first cultivated at Mission San Diego de Alcalá in 1769 or later around 1795. Orchards were started at other missions but in 1838 an inspection found only two olive orchards in California. Cultivation for oil gradually became a highly successful commercial venture from the 1860s onward. In Japan the first successful planting of olive trees happened in 1908 on Shodo Island which became the cradle of olive cultivation. It is estimated that there are about 865 million olive trees in the world today (as of 2005), and the vast majority of these are found in Mediterranean countries, although traditionally marginal areas account for no more than 25% of olive planted area and 10% of oil production.

Symbolic Connotations

Olive oil has long been considered sacred. The olive branch was often a symbol of abundance, glory and peace. The leafy branches of the olive tree were ritually offered to deities and powerful figures as emblems of benediction and purification, and they were used to crown the victors of friendly games and bloody wars. Today, olive oil is still used in many religious ceremonies. Over the years, the olive has been the symbol of peace, wisdom, glory, fertility, power and purity.

Ancient Egypt

Leafy branches of the olive tree were found in Tutankhamun's tomb.

Ancient Israel and Hebrew Bible

The olive was one of the main elements in ancient Israelite cuisine. Olive oil was used for not only food and cooking, but also lighting, sacrificial offerings, ointment, and anointment for priestly or royal office.

The olive tree is one of the first plants mentioned in the Hebrew Bible and in the Christian Old Testament, and one of the most significant. It was an olive leaf that a dove brought back to Noah to demonstrate that the flood was over (Book of Genesis, 8:11). The olive is listed in Deuteronomy 8:8 as one of the seven species that are noteworthy products of the Land of Israel.

Ancient Greece

The ancient Greeks used to smear olive oil on their bodies and hair as a matter of grooming and good health.

Olive oil was used to anoint kings and athletes in ancient Greece. It was burnt in the sacred lamps of temples as well as being the "eternal flame" of the original Olympic Games. Victors in these games were crowned with its leaves.

In Homer's *Odyssey*, Odysseus crawls beneath two shoots of olive that grow from a single stock, and in the *Iliad*, (XVII.53ff) is a metaphoric description of a lone olive tree in the mountains, by a spring; the Greeks observed that the olive rarely thrives at a distance from the sea, which in Greece invariably means up mountain slopes. Greek myth attributed to the primordial culture-hero Aristaeus the understanding of olive husbandry, along with cheese-making and bee-keeping. Olive was one of the woods used to fashion the most primitive Greek cult figures, called *xoana*, referring to their wooden material; they were reverently preserved for centuries. It was purely a matter of local pride that the Athenians claimed that the olive grew first in Athens. In an archaic Athenian foundation myth, Athena won the patronship of Attica from Poseidon with the gift of the olive. Though, according to the 4th-century BC father of botany, Theophrastus, olive trees ordinarily attained an age of about 200 years, he mentions that the very olive tree of Athena still grew on the Acropolis; it was still to be seen there in the 2nd century AD; and when Pausanias was shown it, c. 170 AD, he reported "Legend also says that when the Persians fired Athens the olive was burnt down, but on the very day it was burnt it grew again to the height of two cubits." Indeed, olive suckers sprout readily from the stump, and the great age of some existing olive trees shows that it was perfectly possible that the olive tree of the Acropolis dated to the Bronze Age. The olive was sacred to Athena and appeared on the Athenian coinage.

Theophrastus, in *On the Causes of Plants*, does not give as systematic and detailed an account of olive husbandry as he does of the vine, but he makes clear (in 1.16.10) that the cultivated olive must be vegetatively propagated; indeed, the pits give rise to thorny, wild-type olives, spread far and wide by birds. Theophrastus reports how the bearing olive can be grafted on the wild olive, for which the Greeks had a separate name, *kotinos*. In his *Enquiry into Plants* (2.1.2-4) he states that the olive can be propagated from a piece of the trunk, the root, a twig, or a stake.

Ancient Rome

According to Pliny the Elder, a vine, a fig tree and an olive tree grew in the middle of the Roman Forum; the latter was planted to provide shade (the garden plot was recreated in the 20th century). The Roman poet Horace mentions it in reference to his own diet, which he describes as very simple: "As for me, olives, endives, and smooth mallows provide sustenance." Lord Monboddo comments on the olive in 1779 as one of the foods preferred by the ancients and as one of the most perfect foods.

Vitruvius describes of the use of charred olive wood in tying together walls and foundations in his *De Architectura*:

The thickness of the wall should, in my opinion, be such that armed men meeting on top of it may pass one another without interference. In the thickness there should be set a very close succession of ties made of charred olive wood, binding the two faces of the wall together like pins, to give it lasting endurance. For that is a material which neither decay, nor the weather, nor time can harm, but even though buried in the earth or set in the water it keeps sound and useful forever. And so not only city walls but substructures in general and all walls that require a thickness like that of a city wall, will be long in falling to decay if tied in this manner.

Storing olives on Dere Street; Tacuinum Sanitatis, 14th century,

New Testament

The Mount of Olives east of Jerusalem is mentioned several times in the New Testament. The Allegory of the Olive Tree in St. Paul's Epistle to the Romans refers to the scattering and gathering of Israel. It compares the Israelites to a tame olive tree and the Gentiles to a wild olive branch. The olive tree itself, as well as olive oil and olives, play an important role in the Bible.

Islam

The olive tree and olive oil are mentioned seven times in the Quran, and the olive is praised as a precious fruit. Olive tree and olive-oil health benefits have been propounded in Prophetic medicine. Muhammad is reported to have said: "Take oil of olive and massage with it – it is a blessed tree" (Sunan al-Darimi, 69:103).

Olives are substitutes for dates (if not available) during Ramadan fasting, and olive tree leaves are used as incense in some Muslim Mediterranean countries.

Oldest Known Olive Trees

Many olive trees in the groves around the Mediterranean are said to be centuries old, and ages as great as 2000 years have been demonstrated for some individual trees.

An olive tree on the island of Brijuni (Brioni), Istria in Croatia, has a radiocarbon dating age of about 1,600 years. It still gives fruit (about 30 kg or 66 lb per year), which is made into top quality olive oil.

Kaštela, Croatia

Pliny the Elder recorded a story about a sacred Greek olive tree that was 1,600 years old. An olive tree in west Athens, named "Plato's Olive Tree", is supposed to be a remnant of the grove within which Plato's Academy was situated, which would make it approximately 2,400 years old. The tree comprised a cavernous trunk from which a few branches were still sprouting in 1975, when a traffic accident caused a bus to uproot it. Since then, the trunk has been preserved and displayed in the nearby Agricultural University of Athens. A supposedly older tree, the "Peisistratos Tree", is located by the banks of the Cephisus River, in the municipality of Agioi Anargyroi, and is said to be a remnant of an olive grove that was planted by Athenian tyrant Peisistratos in the 6th century BC. Numerous ancient olive trees also exist near Pelion in Greece. The age of an olive tree in Crete, the Finix Olive is claimed to be over 2,000 years old; this estimate is based on archaeological evidence around the tree. The Olive tree of Vouves, also in Crete, has an age estimated between 2000 and 4000 years. An olive tree called *Farga d'Arió* in Ulldecona, Catalonia, Spain, has been dated (with laser-perimetry methods) as being 1,701 years old, namely it was planted when Constantine the Great was Roman Emperor.

Bar, Montenegro

Canneto Sabino, Italy

Some Italian olive trees are believed to date back to Roman times, although identifying progenitor trees in ancient sources is difficult. A tree located in Santu Baltolu di Carana, in the municipality of Luras in Sardinia, Italy, is respectfully named in Sardinian as the *Ozzastru* by the islanders, and is claimed to be between 3,000 and 4,000 years old according to different studies. There are several other trees of about 1,000 years old within the same garden. The 15th-century trees of *Olivo della Linza*, at Alliste in the Province of Lecce in Apulia on the Italian mainland, were noted by Bishop Ludovico de Pennis during his pastoral visit to the Diocese of Nardò-Gallipoli in 1452.

The town of Bshaale, Lebanon claims to have the oldest olive trees in the world (4000 BC for the oldest), but no scientific study supports these claims. Other trees in the towns of Amioun appear to be at least 1,500 years old.

There are dozens of ancient olive trees throughout Israel and Palestine whose age has earlier been estimated to be 1,600–2,000 years old; however, these estimates could not be supported by current scientific practices. Ancient trees include two giant olive trees in Arraba and five trees in Deir Hanna, both in the Galilee region, which have been determined to be over 3,000 years old, although there is no available data to support the credibility of the study that produced these age estimates and as such the 3000 years age estimate can not be considered valid. All seven trees continue to produce olives.

Several trees in the Garden of Gethsemane (from the Hebrew words "gat shemanim" or olive press) in Jerusalem are claimed to date back to the purported time of Jesus. A study conducted by the National Research Council of Italy in 2012 used carbon dating on older parts of the trunks of three trees from Gethsemane and came up with the dates of 1092, 1166 and 1198 AD, while DNA tests show that the trees were originally planted from the same parent plant. According to molecular analysis, the tested trees showed the same allelic profile at all microsatellite loci analyzed which furthermore may indicate attempt to keep the linage of an older species intact. However Bernabei writes, "All the tree trunks are hollow inside so that the central, older wood is missing . . . In the end, only three from a total of eight olive trees could be successfully dated. The dated ancient olive trees do, however, not allow any hypothesis to be made with regard to the age of the remaining five giant olive trees." Babcox concludes, "The roots of the eight oldest trees are possibly much older. Visiting guides to the garden often state that they are two thousand years old."

The 2,000-year-old Bidni olive trees, which have been confirmed through carbon dating, have been protected since 1933, and are also listed in UNESCO's Database of National Cultural Heritage Laws. In 2011, after recognising their historical and landscape value, and in recognition of the fact that "only 20 trees remain from 40 at the beginning of the 20th century", Maltese authorities declared the ancient Bidni olive grove at Bidnija, limits of Mosta, as a Tree Protected Area, in accordance with the provisions of the Trees and Woodlands Protection Regulations, 2011, as per Government Notice number 473/11.

An olive tree in Bar, Montenegro, is claimed to be over 2,000 years old.

An olive tree in Algarve, Portugal, is 2000 years old, according to radiocarbon dating.

Uses

The olive tree, *Olea europaea*, has been cultivated for olive oil, fine wood, olive leaf, and the olive fruit. 90% of all harvested olives are turned into oil, while about 10% are used as table olives. The olive is one of the "trinity" or "triad" of basic ingredients in Mediterranean cuisine, the other two being wheat for bread, pasta and couscous, and the grape for wine.

green olives

black olives

Table Olives

Table olives are classified by the IOC into 3 groups according to the degree of ripeness achieved before harvesting:

1. Green olives. Picked when they have obtained full size, but before the ripening cycle has begun. Usually shades of green to yellow.

2. Semi-ripe or turning-colour olives. Picked at the beginning of the ripening cycle, when the colour has begun to change from green to multi-colour shades of red to brown. Only the skin is coloured as the flesh of the fruit lacks pigmentation at this stage, unlike that of ripe olives.

3. Black olives or ripe olives. Picked at full maturity when fully ripe. Found in assorted shades of purple to brown to black.

Traditional Fermentation and Curing

An olive vat room used for curing

Raw or fresh olives are naturally very bitter; to make them palatable, olives must be cured and fermented, thereby removing oleuropein, a bitter phenolic compound that can reach levels of 14% of dry matter in young olives. In addition to oleuropein, other phenolic compounds render freshly picked olives unpalatable and must also be removed or lowered in quantity through curing and fermentation. Generally speaking, phenolics reach their peak in young fruit and are converted as the fruit matures. (One exception is the throubes olive, which can be eaten fresh.) Once ripening occurs the levels of phenolics sharply decline through their conversion to other organic products which renders some cultivars edible immediately.

The curing process may take from a few days, with lye, to a few months with brine or salt packing. With the exception of California style and salt cured olives, all methods of curing involve a major fermentation involving bacteria and yeast that is of equal importance to the final table olive product. Traditional cures, using the natural microflora on the fruit to induce fermentation, lead to two important outcomes: the leaching out and breakdown of oleuropein and other unpalatable phenolic compounds, and the generation of favourable metabolites from bacteria and yeast, such as organic acids, probiotics, glycerol and esters, which affect the sensorial properties of the final table olives. The probiotic qualities of mixed bacterial/yeast olive fermentations are only recently being explored. Of all the metabolites, lactic acid is the most important as it acts as a natural preservative lowering the pH of the solution to make the final product more stable against the growth of unwanted pathogenic species. The result is table olives which will store with or without refrigeration, and thus lactic acid bacteria (LAB) dominated fermentations are generally considered the most suitable method of curing olives. Yeast dominated fermentations produce a different suite of metabolites which have fewer self-preservation characteristics, and therefore are acid corrected, often with citric acid, in the final processing stage to achieve microbial stability.

There are many types of preparations for table olives depending on local tastes and traditions. The most important commercial examples are:

Spanish or Sevillian type (Olives with fermentation). Most commonly applied to green olive preparation. Around 60% of all the world's table olives are produced with this method. Olives are soaked in lye (dilute NaOH, 2-4%) for 8–10 hours to hydrolyse the oleuropein. They are usually considered "treated" when the lye has penetrated two-thirds of the way into the fruit. They are then washed once or several times in water to remove the caustic solution and transferred to fermenting vessels full of brine at typical concentrations of 8-12% NaCl. The brine is changed on a regular basis to help remove the phenolic compounds. Fermentation is carried out by the natural microbiota present on the olives that survive the lye treatment process. Many organisms are involved, usually reflecting the local conditions or "Terroir" of the olives. During a typical fermentation gram-negative enterobacteria flourish in small numbers at first, but are rapidly outgrown by lactic acid bacteria species such as *Leuconostoc mesenteroides, Lactobacillus plantarum, Lactobacillus brevis* and *Pediococcus damnosus.* These bacteria produce lactic acid to help lower the pH of the brine and therefore stabilize the product against unwanted pathogenic species. A diversity of yeasts then accumulate in sufficient numbers to help complete the fermentation alongside the lactic acid bacteria. Yeasts commonly mentioned include the teleomorphs *Pichia anomala, Pichia membranifaciens, Debaryomyces hansenii* and *Kluyveromyces marxianus.* Once fermented, the olives are placed in fresh brine and acid corrected, to be ready for market.

Sicilian or Greek type. (Olives with fermentation). Applied to green, semi-ripe and ripe olives. Almost identical to the Spanish type fermentation process, however the lye treatment process is skipped and the olives are placed directly in fermentation vessels full of brine (8-12% NaCl). The brine is changed on a regular basis to help remove the phenolic compounds. As the caustic treatment is avoided, lactic acid bacteria are only present in similar numbers to yeast and appear to be outcompeted by the abundant yeasts found on untreated olives. As there is very little acid produced by the yeast fermentation, lactic, acetic, or citric acid is often added to the fermentation stage to stabilize the process.

Picholine or directly-brined type. (Olives with fermentation). Can be applied to green, semi-ripe or ripe olives. Olives are soaked in lye typically for longer periods than Spanish style (e.g. 10–72 hours) until the solution has penetrated three-quarters of the way into the fruit. They are then washed and immediately brined and acid corrected with citric acid to achieve microbial stability. Fermentation still occurs carried out by acidogenic yeast and bacteria, but is more subdued than other methods. The brine is changed on a regular basis to help remove the phenolic compounds and a series of progressively stronger concentrations of NaCl are added until the product is fully stabilized and ready to be eaten.

Water-cured type. (Olives with fermentation). Can be applied to green, semi-ripe or ripe olives. Olives are soaked in water or weak brine and this solution is changed on a daily basis for 10–14 days. The oleuropein is naturally dissolved and leached into the water and removed during a continual soak-wash cycle. Fermentation takes place during the water treatment stage and involves a mixed yeast/bacteria ecosystem. Sometimes, the olives are lightly cracked with a hammer or a stone to trigger fermentation and speed up the fermentation process. Once debittered the olives are brined to concentrations of 8-12% NaCl and acid corrected, and are then ready to eat.

Salt-cured type. (Olives with minor fermentation). Applied only to ripe olives and usually produced in Morocco or Turkey and other eastern Mediterranean countries. Once picked, the olives are vigorously washed and packed in alternating layers with salt. The high concentrations of salt draw the moisture out of olives, dehydrating and shriveling them until they look somewhat analogous to a raisin. Once packed in salt, fermentation is minimal and only initiated by the most halophilic yeast species such as *Debaryomyces hansenii*. Once cured, they are sold in their natural state without any additives.

California or "artificial ripening" type. (Olives without fermentation). Applied to green and semi-ripe olives. Olives are placed in lye and soaked. Upon their removal they are washed in water injected with compressed air. This process is repeated several times until both oxygen and lye have soaked through to the pit. The repeated, saturated exposure to air oxidises the skin and flesh of the fruit, turning it black in an artificial process that mimics natural ripening. Once fully oxidised or "blackened", they are brined and acid corrected and are then ready for eating.

Olive Wood

Olive wood is very hard and is prized for its durability, colour, high combustion temperature and interesting grain patterns. Because of the commercial importance of the fruit, and the slow growth and relatively small size of the tree, olive wood and its products are relatively expensive. Common uses of the wood include: kitchen utensils, carved wooden bowls, cutting boards, fine furniture, and decorative items.

The yellow or light greenish-brown wood is often finely veined with a darker tint; being very hard and close-grained, it is valued by woodworkers.

Cultivation

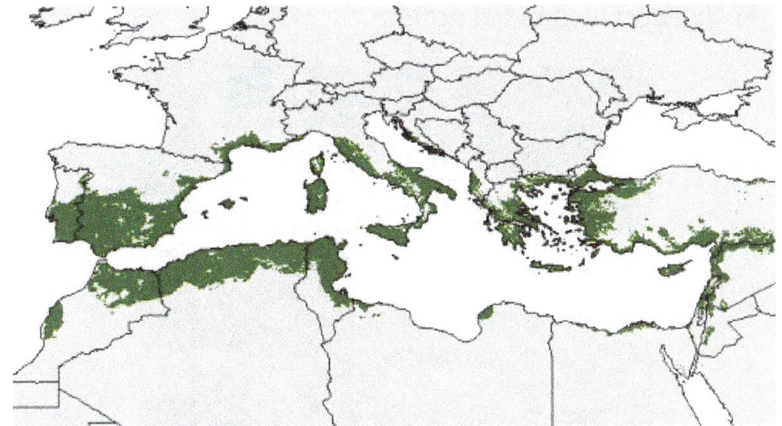

Potential distribution of olive tree over the Mediterranean Basin (Oteros, 2014)

The earliest evidence for the domestication of olives comes from the Chalcolithic Period archaeological site of Teleilat Ghassul in what is today modern Jordan. Farmers in ancient times believed that olive trees would not grow well if planted more than a certain distance from the sea; Theophrastus gives 300 stadia (55.6 km or 34.5 mi) as the limit. Modern experience does not always confirm this, and, though showing a preference for the coast, they have long been grown further inland in some areas with suitable climates, particularly in the southwestern Mediterranean (Iberia, northwest Africa) where winters are mild.

Olive plantation in Andalucía, Spain

Olives are now cultivated in many regions of the world with Mediterranean climates, such as South Africa, Chile, Peru, Australia, and California and in areas with temperate climates such as New Zealand, under irrigation in the Cuyo region in Argentina which has a desert climate. They are also grown in the Córdoba Province, Argentina, which has a temperate climate with rainy summers and dry winters (Cwa). The climate in Argentina changes the external characteristics of the plant but the fruit keeps its original features. The northernmost olive grove is placed in Anglesey, an island off the north west coast of Wales, in the United Kingdom: but it is too early to say if the growing will be successful, having been planted in 2006.

Olives at a market in Toulon, France

Growth and Propagation

Olive trees on Thassos, Greece

Olive trees, *Olea europaea,* show a marked preference for calcareous soils, flourishing best on limestone slopes and crags, and coastal climate conditions. They grow in any light soil, even on clay if well drained, but in rich soils they are predisposed to disease and produce poorer oil than in poorer soil. (This was noted by Pliny the Elder.) Olives like hot weather and sunny positions without any shade while temperatures below −10 °C (14 °F) may injure even a mature tree. They tolerate drought well, thanks to their sturdy and extensive root system. Olive trees can live for several centuries and can remain productive for as long if they are pruned correctly and regularly.

There are only a handful of olive varieties that can be used to cross-pollinate. Pendolino olive trees are partially self-fertile, but pollenizers are needed for a large fruit crop. Other compatible olive tree pollenizers include Leccino and Maurino. Pendolino olive trees are used extensively as pollenizers in large olive tree groves.

Olives are propagated by various methods. The preferred ways are cuttings and layers; the tree roots easily in favourable soil and throws up suckers from the stump when cut down. However, yields from trees grown from suckers or seeds are poor; they must be budded or grafted onto other specimens to do well (Lewington and Parker, 114). Branches of various thickness cut into lengths of about 1 m (3.3 ft) planted deeply in manured ground soon vegetate. Shorter pieces are sometimes laid horizontally in shallow trenches and, when covered with a few centimetres of soil, rapidly throw up sucker-like shoots. In Greece, grafting the cultivated tree on the wild tree is a common practice. In Italy, embryonic buds, which form small swellings on the stems, are carefully excised and planted under the soil surface, where they soon form a vigorous shoot.

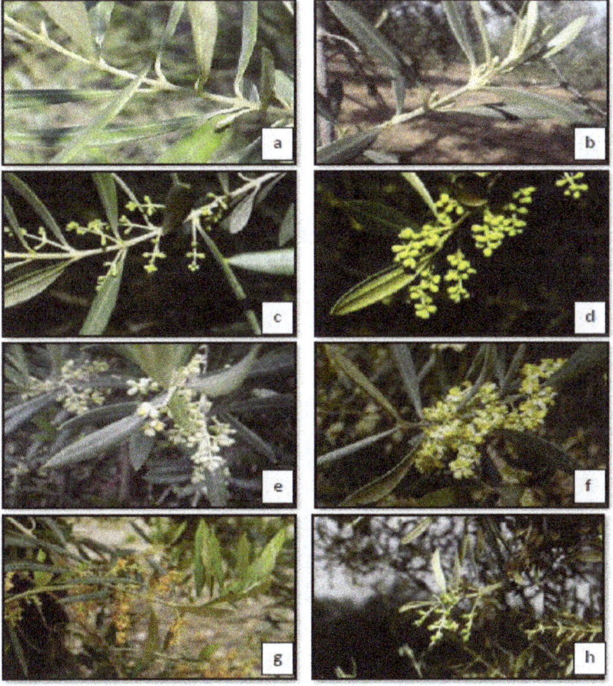

Phenological development of Olive flowering, following BBCH standard scale. a-50, b-51, c-54, d-57, (<15% open flowers); f-65, (>15% open flowers); g-67, (<15% open flowers); h-68 (Oteros et al., 2013)

The olive is also sometimes grown from seed. To facilitate germination, the oily pericarp is first softened by slight rotting, or soaked in hot water or in an alkaline solution.

In situations where extreme cold has damaged or killed the olive tree the rootstock can survive and produce new shoots which in turn become new trees. In this way olive trees can regenerate themselves. In Tuscany in 1985 a very severe frost destroyed many productive, and aged, olive trees and ruined many farmers' livelihoods. However new shoots appeared in the spring and, once the dead wood was removed, became the basis for new fruit-producing trees. In this way an olive tree can live for centuries or even millennia.

Olives grow very slowly, and over many years the trunk can attain a considerable diameter. A. P. de Candolle recorded one exceeding 10 m (33 ft) in girth. The trees rarely exceed 15 m (49 ft) in height, and are generally confined to much more limited dimensions by frequent pruning.

The olive tree, *Olea europaea*, is very hardy: drought-, disease- and fire-resistant, it can live to a great age. Its root system is robust and capable of regenerating the tree even if the above-ground structure is destroyed. The older the olive tree, the broader and more gnarled the trunk becomes. Many olive trees in the groves around the Mediterranean are said to be hundreds of years old, while an age of 2,000 years is claimed for a number of individual trees; in some cases, this has been scientifically verified.

The crop from old trees is sometimes enormous, but they seldom bear well two years in succession, and in many cases a large harvest occurs every sixth or seventh season.

Where the olive is carefully cultivated, as in Languedoc and Provence, the trees are regularly pruned. The pruning preserves the flower-bearing shoots of the preceding year, while keeping the tree low enough to allow the easy gathering of the fruit.

The spaces between the trees are regularly fertilized.

Pests, Diseases, and Weather

There are various pathologies that can affect olives. The most serious pest is the olive fruit fly (*Dacus oleae* or *Bactrocera oleae*) which lays its eggs in the olive most commonly just before it becomes ripe in the autumn. The region surrounding the puncture rots, becomes brown and takes a bitter taste making the olive unfit for eating or for oil. For controlling the pest the practice has been to spray with insecticides (organophosphates, e.g. dimethoate). Classic organic methods have now been applied such as trapping, applying the bacterium Bacillus thuringiensis and spraying with kaolin. Such methods are obligatory for organic olives.

A fungus, *Cycloconium oleaginum*, can infect the trees for several successive seasons, causing great damage to plantations. A species of bacterium, *Pseudomonas savastanoi* pv. *oleae*, induces tumour growth in the shoots. Certain lepidopterous caterpillars feed on the leaves and flowers.

A pest which spreads through olive trees is the black scale bug, a small black scale insect that resembles a small black spot. They attach themselves firmly to olive trees and reduce the quality of the fruit; their main predators are wasps. The curculio beetle eats the edges of leaves, leaving sawtooth damage.

Rabbits eat the bark of olive trees and can do considerable damage, especially to young trees. If the bark is removed around the entire circumference of a tree it is likely to die. Voles and mice also do damage by eating the roots of olives.

At the northern edge of their cultivation zone, for instance in Southern France and north-central Italy, olive trees suffer occasionally from frost. Gales and long-continued rains during the gathering season also cause damage.

As an Invasive Species

Since its first domestication, *Olea europaea* has been spreading back to the wild from planted groves. Its original wild populations in southern Europe have been largely swamped by feral plants.

In some other parts of the world where it has been introduced, most notably South Australia, the olive has become a major woody weed that displaces native vegetation. In South Australia, its seeds are spread by the introduced red fox and by many bird species, including the European starling and the native emu, into woodlands, where they germinate and eventually form a dense canopy that prevents regeneration of native trees. As the climate of South Australia is very dry and bushfire prone, the oil rich feral olive tree substantially increases the fire hazard of native sclerophyll woodlands.

Olives as invasive weeds, Adelaide Hills, Australia

Harvest and Processing

Olives are harvested in the autumn and winter. More specifically in the Northern hemisphere, green olives are picked from the end of September to about the middle of November. Blond olives are picked from the middle of October to the end of November, and black olives are collected from the middle of November to the end of January or early February. In southern Europe, harvesting is done for several weeks in winter, but the time varies in each country, and with the season and the cultivar.

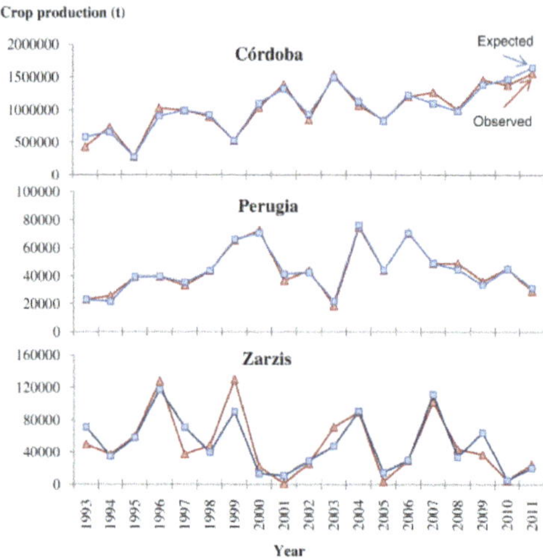

Forecasting olive crop production based on aerobiological method (Oteros et al., 2014)

Most olives today are harvested by shaking the boughs or the whole tree. Using olives found lying on the ground can result in poor quality oil, due to damage. Another method involves standing on a ladder and "milking" the olives into a sack tied around the harvester's waist. This method produces high quality oil. A third method uses a device called an oli-net that wraps around the tree trunk and opens to form an umbrella-like catcher from which workers collect the fruit. Another method uses an electric tool, 'the oliviera', that has large tongs that spin around quickly, removing fruit from the tree. Olives harvested by this method are used for oil.

Table olive varieties are more difficult to harvest, as workers must take care not to damage the fruit; baskets that hang around the worker's neck are used. In some places in Italy, Croatia, and Greece, olives are harvested by hand because the terrain is too mountainous for machines. As a result, the fruit is not bruised, which leads to a superior finished product. The method also involves sawing off branches, which is healthy for future production.

The amount of oil contained in the fruit differs greatly by cultivar; the pericarp is usually 60–70% oil. Typical yields are 1.5–2.2 kg (3.3–4.9 lb) of oil per tree per year.

Global Production

Olives are one of the most extensively cultivated fruit crops in the world. In 2011 there were about 9.6 million hectares planted with olive trees, which is more than twice the amount of land devoted to apples, bananas or mangoes. Only coconut trees and oil palms command more space. Cultivation area tripled from 2,600,000 to 7,950,000 hectares (6,400,000 to 19,600,000 acres) between 1960 and 1998 and reached a 10 million ha peak in 2008. The ten largest producing countries, according to the Food and Agriculture Organization, are all located in the Mediterranean region and produce 95% of the world's olives.

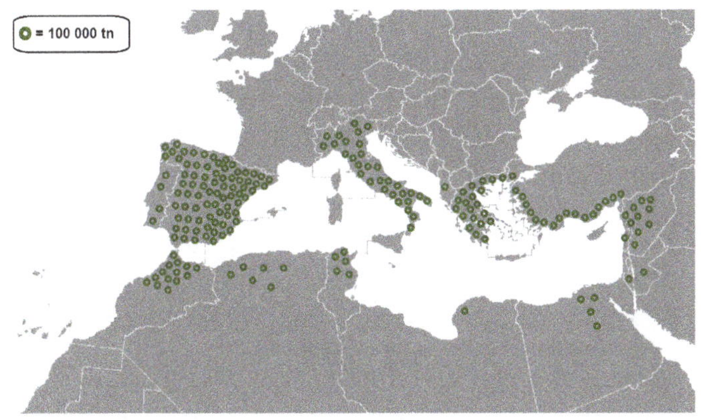

Main countries of production (Year 2011 per FAOSTAT)				
Rank	**Country/Region**	**Production (in 1000s tonnes)**	**Cultivated area (in 1000s hectares)**	**Yield (q/ Ha)**
—	World	19,894	9,635	20.598
01	Spain	7,869	2,330	29.781
02	Italy	3,182	1,144	27.806
03	Greece	2,000	850	23.529
04	Turkey	1,750	799	21.916
05	Morocco	1,416	598	22.839
06	Syria	1,095	684	15.997
07	Algeria	611	295	14.237
08	Tunisia	562	1,780	4.848
09	Egypt	460	53	87.273
10	Portugal	444	343	12.931

Nutrition

One hundred grams of cured green olives provide 146 calories, are a rich source of vitamin E (25% of the Daily Value, DV), and contain a large amount of sodium (104% DV); other nutrients are insignificant. Green olives are 75% water, 15% fat, 4% carbohydrates and 1% protein (table).

The polyphenol composition of olive fruits varies during fruit ripening and during processing by fermentation when olives are immersed whole in brine or crushed to produce oil. In raw fruit, total polyphenol contents, as measured by the Folin method, are 117 mg/100 g in black olives and 161 mg/100 g in green olives, compared to 55 and

21 mg/100 g for extra virgin and virgin olive oil, respectively. Olive fruit contains several types of polyphenols, mainly tyrosols, phenolic acids, flavonols and flavones, and for black olives, anthocyanins. The main bitter flavor of olives before curing results from oleuropein and its aglycone which total in content, respectively, 72 and 82 mg/100 g in black olives, and 56 and 59 mg/100 g in green olives.

During the crushing, kneading and extraction of olive fruit to obtain olive oil, oleuropein, demethyloleuropein and ligstroside are hydrolyzed by endogenous beta-glucosidases to form aldehydic aglycones. The aglycones become soluble in the oil phase, whereas the glycosides remain in the water phase.

Polyphenol content also varies with olive cultivar (Spanish Manzanillo highest) and the manner of presentation, with plain olives having higher contents than those that are pitted or stuffed.

Allergenic Potential

Olive tree pollen is extremely allergenic, with an OPALS allergy scale rating of 10 out of 10. *Olea europaea* is primarily wind-pollinated, and their light, buoyant pollen is a strong trigger for asthma. One popular variety, "Swan Hill", is widely sold as an "allergy-free" olive tree; however, this variety does bloom and produce allergenic pollen.

Cheese

Coulommiers cheese

Cheese is a food derived from milk that is produced in a wide range of flavors, textures, and forms by coagulation of the milk protein casein. It comprises proteins and fat from milk, usually the milk of cows, buffalo, goats, or sheep. During production, the milk is usually acidified, and adding the enzyme rennet causes coagulation. The solids are separated and pressed into final form. Some cheeses have molds on the rind or throughout. Most cheeses melt at cooking temperature.

A platter with cheese and garnishes

Hundreds of types of cheese from various countries are produced. Their styles, textures and flavors depend on the origin of the milk (including the animal's diet), whether they have been pasteurized, the butterfat content, the bacteria and mold, the processing, and aging. Herbs, spices, or wood smoke may be used as flavoring agents. The yellow to red color of many cheeses, such as Red Leicester, is produced by adding annatto. Other ingredients may be added to some cheeses, such as black pepper, garlic, chives or cranberries.

A variety of cheeses

For a few cheeses, the milk is curdled by adding acids such as vinegar or lemon juice. Most cheeses are acidified to a lesser degree by bacteria, which turn milk sugars into lactic acid, then the addition of rennet completes the curdling. Vegetarian alternatives to rennet are available; most are produced by fermentation of the fungus *Mucor miehei*, but others have been extracted from various species of the *Cynara* thistle family. Cheesemakers near a dairy region may benefit from fresher, lower-priced milk, and lower shipping costs.

Cheese is valued for its portability, long life, and high content of fat, protein, calcium, and phosphorus. Cheese is more compact and has a longer shelf life than milk, although how long a cheese will keep depends on the type of cheese; labels on packets of cheese often claim that a cheese should be consumed within three to five days of opening.

Generally speaking, hard cheeses, such as parmesan last longer than soft cheeses, such as Brie or goat's milk cheese. The long storage life of some cheeses, especially when encased in a protective rind, allows selling when markets are favorable.

There is some debate as to the best way to store cheese, but some experts say that wrapping it in cheese paper provides optimal results. Cheese paper is coated in a porous plastic on the inside, and the outside has a layer of wax. This specific combination of plastic on the inside and wax on the outside protects the cheese by allowing condensation on the cheese to be wicked away while preventing moisture from within the cheese escaping.

A specialist seller of cheese is sometimes known as a *cheesemonger*. Becoming an expert in this field requires some formal education and years of tasting and hands-on experience, much like becoming an expert in wine or cuisine. The cheesemonger is responsible for all aspects of the cheese inventory: selecting the cheese menu, purchasing, receiving, storage, and ripening.

Etymology

Cheese on market stand in Basel, Switzerland

The word *cheese* comes from Latin *caseus*, from which the modern word casein is also derived. The earliest source is from the proto-Indo-European root **kwat-*, which means "to ferment, become sour".

More recently, *cheese* comes from *chese* (in Middle English) and *cīese* or *cēse* (in Old English). Similar words are shared by other West Germanic languages—West Frisian *tsiis*, Dutch *kaas*, German *Käse*, Old High German *chāsi*—all from the reconstructed West-Germanic form **kāsī*, which in turn is an early borrowing from Latin.

When the Romans began to make hard cheeses for their legionaries' supplies, a new word started to be used: *formaticum*, from *caseus formatus*, or "molded cheese" (as in "formed", not "moldy"). It is from this word that the French *fromage*, proper Italian *formaggio*, Catalan *formatge*, Breton *fourmaj*, and Provençal *furmo* are derived. Of the Romance languages, Spanish, Portuguese, Romanian, Tuscan and Southern Italian dialects use words derived from *caseus* (*queso*, *queijo*, *caş* and *caso* for example). The word *cheese* itself is occasionally employed in a sense that means "molded" or "formed". *Head cheese* uses the word in this sense.

History

Origins

A piece of soft curd cheese, oven-baked to increase longevity

Cheese is an ancient food whose origins predate recorded history. There is no conclusive evidence indicating where cheesemaking originated, either in Europe, Central Asia or the Middle East, but the practice had spread within Europe prior to Roman times and, according to Pliny the Elder, had become a sophisticated enterprise by the time the Roman Empire came into being.

The earliest evidence of cheese-making in the archaeological record dates back to 5,500 BCE, in what is now Kujawy, Poland, where strainers with milk fats molecules have been found. Earliest proposed dates for the origin of cheesemaking range from around 8000 BCE, when sheep were first domesticated. Since animal skins and inflated internal organs have, since ancient times, provided storage vessels for a range of foodstuffs, it is probable that the process of cheese making was discovered accidentally by storing milk in a container made from the stomach of an animal, resulting in the milk being turned to curd and whey by the rennet from the stomach. There is a legend – with variations – about the discovery of cheese by an Arab trader who used this method of storing milk.

Cheesemaking may have begun independently of this by the pressing and salting of curdled milk to preserve it. Observation that the effect of making cheese in an animal stomach gave more solid and better-textured curds may have led to the deliberate addition of rennet.

Early archeological evidence of Egyptian cheese has been found in Egyptian tomb murals, dating to about 2000 BCE. The earliest cheeses were likely to have been quite sour and salty, similar in texture to rustic cottage cheese or feta, a crumbly, flavorful Greek cheese.

Cheese produced in Europe, where climates are cooler than the Middle East, required less salt for preservation. With less salt and acidity, the cheese became a suitable environment for useful microbes and molds, giving aged cheeses their respective flavors.

The earliest ever discovered preserved cheese was found in the Taklamakan Desert in Xinjiang, China, and it dates back as early as 1615 BCE.

Ancient Greece and Rome

Cheese in a market in Italy

Ancient Greek mythology credited Aristaeus with the discovery of cheese. Homer's *Odyssey* (8th century BCE) describes the Cyclops making and storing sheep's and goats' milk cheese (translation by Samuel Butler):

We soon reached his cave, but he was out shepherding, so we went inside and took stock of all that we could see. His cheese-racks were loaded with cheeses, and he had more lambs and kids than his pens could hold...

When he had so done he sat down and milked his ewes and goats, all in due course, and then let each of them have her own young. He curdled half the milk and set it aside in wicker strainers.

By Roman times, cheese was an everyday food and cheesemaking a mature art. Columella's *De Re Rustica* (circa 65 CE) details a cheesemaking process involving rennet coagulation, pressing of the curd, salting, and aging. Pliny's *Natural History* (77 CE) devotes a chapter (XI, 97) to describing the diversity of cheeses enjoyed by Romans of the early Empire. He stated that the best cheeses came from the villages near Nîmes, but did not keep long and had to be eaten fresh. Cheeses of the Alps and Apennines were as remarkable for their variety then as now. A Ligurian cheese was noted for being made mostly from sheep's milk, and some cheeses produced nearby were stated to weigh as much as a thousand pounds each. Goats' milk cheese was a recent taste in Rome, improved over the "medicinal taste" of Gaul's similar cheeses by smoking. Of cheeses from overseas, Pliny preferred those of Bithynia in Asia Minor.

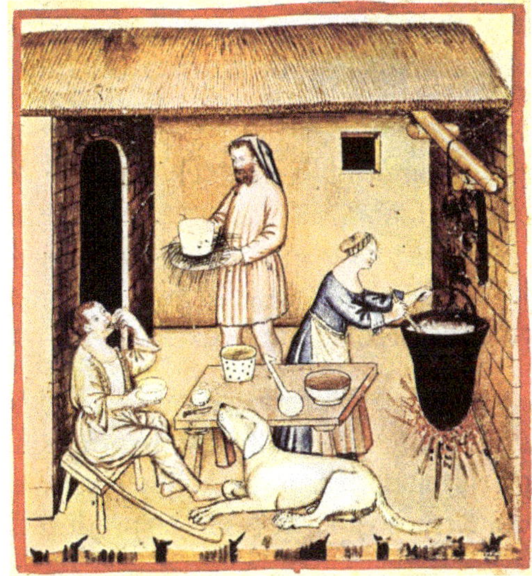

Cheese, Tacuinum sanitatis Casanatensis (14th century)

Post-Roman Europe

As Romanized populations encountered unfamiliar newly settled neighbors, bringing their own cheese-making traditions, their own flocks and their own unrelated words for *cheese*, cheeses in Europe diversified further, with various locales developing their own distinctive traditions and products. As long-distance trade collapsed, only travelers would encounter unfamiliar cheeses: Charlemagne's first encounter with a white cheese that had an edible rind forms one of the constructed anecdotes of Notker's *Life* of the Emperor.

The British Cheese Board claims that Britain has approximately 700 distinct local cheeses; France and Italy have perhaps 400 each. (A French proverb holds there is a different French cheese for every day of the year, and Charles de Gaulle once asked "how can you govern a country in which there are 246 kinds of cheese?") Still, the advancement of the cheese art in Europe was slow during the centuries after Rome's fall. Many cheeses today were first recorded in the late Middle Ages or after—cheeses like Cheddar around 1500, Parmesan in 1597, Gouda in 1697, and Camembert in 1791.

In 1546 *The Proverbs of John Heywood* claimed "the moon is made of a greene cheese." (*Greene* may refer here not to the color, as many now think, but to being new or unaged.) Variations on this sentiment were long repeated and NASA exploited this myth for an April Fools' Day spoof announcement in 2006.

Modern Era

Until its modern spread along with European culture, cheese was nearly unheard of in east Asian cultures, in the pre-Columbian Americas, and only had limited use in sub-Mediterranean Africa, mainly being widespread and popular only in Europe, the Middle East, the Indian subcontinent, and areas influenced by those cultures. But with the spread, first of European imperialism, and later of Euro-American culture and food, cheese has gradually become known and increasingly popular worldwide.

The first factory for the industrial production of cheese opened in Switzerland in 1815, but large-scale production first found real success in the United States. Credit usually goes to Jesse Williams, a dairy farmer from Rome, New York, who in 1851 started making cheese in an assembly-line fashion using the milk from neighboring farms. Within decades, hundreds of such dairy associations existed.

The 1860s saw the beginnings of mass-produced rennet, and by the turn of the century scientists were producing pure microbial cultures. Before then, bacteria in cheesemaking had come from the environment or from recycling an earlier batch's whey; the pure cultures meant a more standardized cheese could be produced.

Factory-made cheese overtook traditional cheesemaking in the World War II era, and factories have been the source of most cheese in America and Europe ever since.

Cheese production – 2013	
Country	**Production (millions of tonnes)**
United States	5.4
Germany	2.2
France	1.9
Italy	1.2
Netherlands	0.8
World	**21.3**
Source: FAOSTAT of the United Nations	

Production

In 2013, world production of cheese was 21.3 million tonnes, with the United States accounting for 25% (5.4 million tonnes) of the world total followed by Germany, France and Italy (table).

During 2015, Germany, France, Netherlands and Italy exported 10-14% of their produced cheese. The United States, the biggest world producer of cheese, is a marginal exporter (5.3% of total), as most of its production is for the domestic market.

Consumption

France, Iceland, Finland, Denmark and Germany were the highest consumers of cheese in 2014, averaging 25 kg (55 lb) per person.

Processing

Curdling

During industrial production of Emmental cheese, the as-yet-undrained curd is broken by rotating mixers.

A required step in cheesemaking is separating the milk into solid curds and liquid whey. Usually this is done by acidifying (souring) the milk and adding rennet. The acidification can be accomplished directly by the addition of an acid, such as vinegar, in a few cases (paneer, queso fresco). More commonly starter bacteria are employed instead which convert milk sugars into lactic acid. The same bacteria (and the enzymes they produce) also play a large role in the eventual flavor of aged cheeses. Most cheeses are made with starter bacteria from the *Lactococcus*, *Lactobacillus*, or *Streptococcus* families. Swiss starter cultures also include *Propionibacter shermani*, which produces carbon dioxide gas bubbles during aging, giving Swiss cheese or Emmental its holes (called "eyes").

Some fresh cheeses are curdled only by acidity, but most cheeses also use rennet. Rennet sets the cheese into a strong and rubbery gel compared to the fragile curds produced by acidic coagulation alone. It also allows curdling at a lower acidity—important because flavor-making bacteria are inhibited in high-acidity environments. In general, softer, smaller, fresher cheeses are curdled with a greater proportion of acid to rennet than harder, larger, longer-aged varieties.

While rennet was traditionally produced via extraction from the inner mucosa of the fourth stomach chamber of slaughtered young, unweaned calves, most rennet used today in cheesemaking is produced recombinantly. The majority of the applied chymosin is retained in the whey and, at most, may be present in cheese in trace quantities. In ripe cheese, the type and provenance of chymosin used in production cannot be determined.

Curd Processing

At this point, the cheese has set into a very moist gel. Some soft cheeses are now essentially complete: they are drained, salted, and packaged. For most of the rest, the curd is cut into small cubes. This allows water to drain from the individual pieces of curd.

Some hard cheeses are then heated to temperatures in the range of 35–55 °C (95–131 °F). This forces more whey from the cut curd. It also changes the taste of the finished cheese, affecting both the bacterial culture and the milk chemistry. Cheeses that are heated to the higher temperatures are usually made with thermophilic starter bacteria that survive this step—either *Lactobacilli* or *Streptococci.*

Salt has roles in cheese besides adding a salty flavor. It preserves cheese from spoiling, draws moisture from the curd, and firms cheese's texture in an interaction with its proteins. Some cheeses are salted from the outside with dry salt or brine washes. Most cheeses have the salt mixed directly into the curds.

Cheese factory in the Netherlands

Other techniques influence a cheese's texture and flavor. Some examples are :

- Stretching: (Mozzarella, Provolone) The curd is stretched and kneaded in hot water, developing a stringy, fibrous body.
- Cheddaring: (Cheddar, other English cheeses) The cut curd is repeatedly piled up, pushing more moisture away. The curd is also mixed (or *milled*) for a long time, taking the sharp edges off the cut curd pieces and influencing the final product's texture.
- Washing: (Edam, Gouda, Colby) The curd is washed in warm water, lowering its acidity and making for a milder-tasting cheese.

Most cheeses achieve their final shape when the curds are pressed into a mold or form. The harder the cheese, the more pressure is applied. The pressure drives out moisture—the molds are designed to allow water to escape—and unifies the curds into a single solid body.

Parmigiano-Reggiano in a modern factory

Ripening

A newborn cheese is usually salty yet bland in flavor and, for harder varieties, rubbery in texture. These qualities are sometimes enjoyed—cheese curds are eaten on their own—but normally cheeses are left to rest under controlled conditions. This aging period (also called ripening, or, from the French, *affinage*) lasts from a few days to several years. As a cheese ages, microbes and enzymes transform texture and intensify flavor. This transformation is largely a result of the breakdown of casein proteins and milkfat into a complex mix of amino acids, amines, and fatty acids.

Some cheeses have additional bacteria or molds intentionally introduced before or during aging. In traditional cheesemaking, these microbes might be already present in the aging room; they are simply allowed to settle and grow on the stored cheeses. More often today, prepared cultures are used, giving more consistent results and putting fewer constraints on the environment where the cheese ages. These cheeses include soft ripened cheeses such as Brie and Camembert, blue cheeses such as Roquefort, Stilton, Gorgonzola, and rind-washed cheeses such as Limburger.

Types

Feta from Greece

Local cheese at an open-air market in Peru.

There are many types of cheese, with around 500 different varieties recognized by the International Dairy Federation, more than 400 identified by Walter and Hargrove, more than 500 by Burkhalter, and more than 1,000 by Sandine and Elliker. The varieties may be grouped or classified into types according to criteria such as length of ageing, texture, methods of making, fat content, animal milk, country or region of origin, etc.—with these criteria either being used singly or in combination, but with no single method being universally used. The method most commonly and traditionally used is based on moisture content, which is then further discriminated by fat content and curing or ripening methods. Some attempts have been made to rationalise the classification of cheese—a scheme was proposed by Pieter Walstra which uses the primary and secondary starter combined with moisture content, and Walter and Hargrove suggested classifying by production methods which produces 18 types, which are then further grouped by moisture content.

Moisture content (soft to hard)

Categorizing cheeses by firmness is a common but inexact practice. The lines between "soft", "semi-soft", "semi-hard", and "hard" are arbitrary, and many types of cheese are made in softer or firmer variations. The main factor that controls cheese hardness is moisture content, which depends largely on the pressure with which it is packed into molds, and on aging time.

Fresh, whey and stretched curd cheeses

The main factor in the categorization of these cheeses is their age. Fresh cheeses without additional preservatives can spoil in a matter of days.

Content (double cream, goat, ewe and water buffalo)

Some cheeses are categorized by the source of the milk used to produce them or by the added fat content of the milk from which they are produced. While most of the world's commercially available cheese is made from cows' milk, many parts of the world also produce cheese from goats and sheep. Double cream cheeses are soft cheeses of cows' milk enriched with cream so that their fat content is 60% or, in the case of triple creams, 75%. The use of the terms "double" or "triple" is not meant to give a quantitative reference to the change in fat content, since the fat content of whole cows' milk is 3%-4%.

Emmental

Soft-ripened and blue-vein

There are at least three main categories of cheese in which the presence of mold is a significant feature: soft ripened cheeses, washed rind cheeses and blue cheeses.

Processed cheeses

Processed cheese is made from traditional cheese and emulsifying salts, often with the addition of milk, more salt, preservatives, and food coloring. It is inexpensive, consistent, and melts smoothly. It is sold packaged and either pre-sliced or unsliced, in a number of varieties. It is also available in aerosol cans in some countries.

Cooking and Eating

Zigerbrüt, cheese grated onto bread through a mill, from the Canton of Glarus in Switzerland.

Saganaki, lit on fire, served in Chicago.

At refrigerator temperatures, the fat in a piece of cheese is as hard as unsoftened butter, and its protein structure is stiff as well. Flavor and odor compounds are less easily liberated when cold. For improvements in flavor and texture, it is widely advised that cheeses be allowed to warm up to room temperature before eating. If the cheese is further warmed, to 26–32 °C (79–90 °F), the fats will begin to "sweat out" as they go beyond soft to fully liquid.

Above room temperatures, most hard cheeses melt. Rennet-curdled cheeses have a gel-like protein matrix that is broken down by heat. When enough protein bonds are broken, the cheese itself turns from a solid to a viscous liquid. Soft, high-moisture cheeses will melt at around 55 °C (131 °F), while hard, low-moisture cheeses such as Parmesan remain solid until they reach about 82 °C (180 °F). Acid-set cheeses, including halloumi, paneer, some whey cheeses and many varieties of fresh goat cheese, have a protein structure that remains intact at high temperatures. When cooked, these cheeses just get firmer as water evaporates.

Some cheeses, like raclette, melt smoothly; many tend to become stringy or suffer from a separation of their fats. Many of these can be coaxed into melting smoothly in the presence of acids or starch. Fondue, with wine providing the acidity, is a good example of a smoothly melted cheese dish. Elastic stringiness is a quality that is sometimes enjoyed, in dishes including pizza and Welsh rarebit. Even a melted cheese eventually turns solid again, after enough moisture is cooked off. The saying "you can't melt cheese twice" (meaning "some things can only be done once") refers to the fact that oils leach out during the first melting and are gone, leaving the non-meltable solids behind.

As its temperature continues to rise, cheese will brown and eventually burn. Browned, partially burned cheese has a particular distinct flavor of its own and is frequently used in cooking (e.g., sprinkling atop items before baking them).

Nutrition and Health

The nutritional value of cheese varies widely. Cottage cheese may consist of 4% fat and 11% protein while some whey cheeses are 15% fat and 11% protein, and triple-crème cheeses are 36% fat and 7% protein. In general, cheese is a rich source (20% or more of the Daily Value, DV) of calcium, protein, phosphorus, sodium and saturated fat. A 28-gram (0.99 oz) (one ounce) serving of cheddar cheese contains about 7 grams (0.25 oz) of protein and 202 milligrams of calcium. Nutritionally, cheese is essentially concentrated milk: it takes about 200 grams (7.1 oz) of milk to provide that much protein, and 150 grams (5.3 oz) to equal the calcium.

Mozzarella

MacroNutrients (grams) of common cheeses per 100gm

Cheese	Water	Protein	Fat	Carbs
Swiss	37.1	26.9	27.8	5.4
Feta	55.2	14.2	21.3	4.1
Cheddar	36.8	24.9	33.1	1.3
Mozarella	50	22.2	22.4	2.2
Cottage	80	11.1	4.3	3.4

Vitamin contents in %DV of common cheeses per 100gm

Cheese	A	B1	B2	B3	B5	B6	B9	B12	Ch.	C	D	E	K
Swiss	17	4	17	0	4	4	1	**56**	2.8	0	11	2	3
Feta	8	10	50	5	10	21	8	28	2.2	0	0	1	2
Cheddar	20	2	22	0	4	4	5	14	3	0	3	1	3
Mozzarella	14	2	17	1	1	2	2	38	2.8	0	0	1	3
Cottage	3	2	10	0	6	2	3	7	3.3	0	0	0	0

Mineral contents in %DV of common cheeses per 100 grams

Cheese	Ca	Fe	Mg	P	K	Na	Zn	Cu	Mn	Se
Swiss	**79**	10	1	**57**	2	8	29	2	0	26
Feta	**49**	4	5	**34**	2	46	19	2	1	21
Cheddar	72	4	7	51	3	26	21	2	1	20
Mozzarella	**51**	2	5	**35**	2	26	19	1	1	24
Cottage	8	0	2	16	3	15	3	1	0	14

Ch. = Choline; Ca = Calcium; Fe = Iron; Mg = Magnesium; P = Phosphorus; K = Potassium; Na = Sodium; Zn = Zinc; Cu = Copper; Mn = Manganese; Se = Selenium;

Note : All nutrient values including protein are in %DV per 100 grams of the food item except for Macronutrients. Source : Nutritiondata.self.com

Neonatal Infection and Death

Cheese has the potential for promoting the growth of *Listeria* bacteria. *Listeria monocytogenes* can also cause serious infection in an infant and pregnant woman and can be transmitted to her infant in utero or after birth. The infection has the potential of seriously harming or even causing the death of a preterm infant, an infant of low or very low birth weight, or an infant with an immune system deficiency or a congenital defect of the immune system. The presence of this pathogen can sometimes be determined by the symptoms that appear as a gastrointestinal illness in the mother. The mother can also acquire infection from ingesting food that contains other animal products such as, unpasteurized milk, delicatessen meats, and hot dogs.

Heart Disease

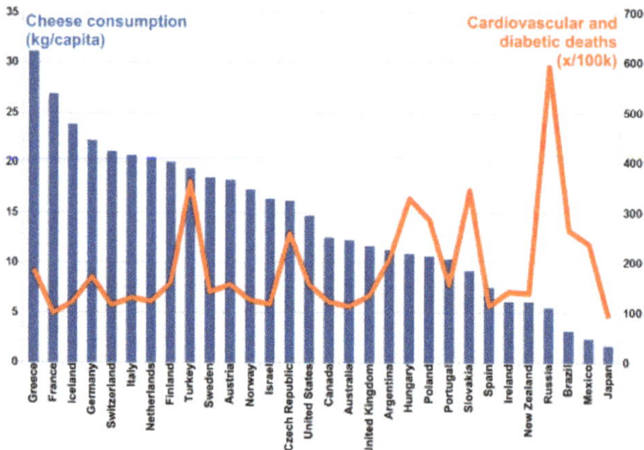

Average cheese consumption and rates of mortality due to cardiovascular disease or diabetes

A review of the medical literature published in 2012 noted that: "Cheese consumption is the leading contributor of SF (saturated fat) in the U.S. diet, and therefore would be predicted to increase LDL-C (LDL cholesterol) and consequently increase the risk of CVD (cardiovascular disease)." It found that: "Based on results from numerous prospective observational studies and meta-analyses, most, but not all, have shown no association and in some cases an inverse relationship between the intake of milk fat containing dairy products and the risk of CVD, CHD (coronary heart disease), and stroke. A limited number of prospective cohort studies found no significant association between the intake of total full-fat dairy products and the risk of CHD or stroke....Most clinical studies showed that full-fat natural cheese, a highly fermented product, significantly lowers LDL-C compared with butter intake of equal total fat and saturated fat content."

Pasteurization

A number of food safety agencies around the world have warned of the risks of raw-milk cheeses. The U.S. Food and Drug Administration states that soft raw-milk cheeses can cause "serious infectious diseases including listeriosis, brucellosis, salmonellosis and tuberculosis". It is U.S. law since 1944 that all raw-milk cheeses (including imports since 1951) must be aged at least 60 days. Australia has a wide ban on raw-milk cheeses as well, though in recent years exceptions have been made for Swiss Gruyère, Emmental and Sbrinz, and for French Roquefort. There is a trend for cheeses to be pasteurized even when not required by law.

Pregnant women may face an additional risk from cheese; the U.S. Centers for Disease Control has warned pregnant women against eating soft-ripened cheeses and blue-veined cheeses, due to the listeria risk, which can cause miscarriage or harm the fetus.

Cultural Attitudes

A cheese merchant in a French market

Although cheese is a vital source of nutrition in many regions of the world and is extensively consumed in others, its use is not universal.

Cheese is rarely found in East Asian cuisines, presumably for historical reasons However, East Asian sentiment against cheese is not universal. In Nepal, the Dairy Development Corporation commercially manufactures cheese made from yak milk and a hard cheese made from either cow or yak milk knows as chhurpi. The national dish of Bhutan, *ema datsi*, is made from homemade yak or mare milk cheese and hot peppers. In Yunnan China, several ethnic minority groups produce Rushan and Rubing from cow's milk. Cheese consumption may be increasing in China, with annual sales doubling from 1996 to 2003 (to a still small 30 million U.S. dollars a year). Certain kinds of Chinese preserved bean curd are sometimes misleadingly referred to in English as "Chinese cheese" because of their texture and strong flavor.

Strict followers of the dietary laws of Islam and Judaism must avoid cheeses made with rennet from animals not slaughtered in a manner adhering to halal or kosher laws. Both faiths allow cheese made with vegetable-based rennet or with rennet made from animals that were processed in a halal or kosher manner. Many less orthodox Jews also believe that rennet undergoes enough processing to change its nature entirely and do not consider it to ever violate kosher law. As cheese is a dairy food, under kosher rules it cannot be eaten in the same meal with any meat.

A traditional Polish sheep's cheese market in Zakopane, Poland

Rennet derived from animal slaughter, and thus cheese made with animal-derived rennet, is not vegetarian. Most widely available vegetarian cheeses are made using rennet produced by fermentation of the fungus *Mucor miehei.* Vegans and other dairy-avoiding vegetarians do not eat conventional cheese, but some vegetable-based cheese substitutes (soy or almond) are used as substitutes.

Even in cultures with long cheese traditions, consumers may perceive some cheeses that are especially pungent-smelling or mold-bearing varieties such as Limburger or Roquefort, as unpalatable. Such cheeses are an acquired taste because they are processed using molds or microbiological cultures, allowing odor and flavor molecules to resemble those in rotten foods. One author stated: "An aversion to the odor of decay has the obvious biological value of steering us away from possible food poisoning, so it is no wonder that an animal food that gives off whiffs of shoes and soil and the stable takes some getting used to."

Collecting cheese labels is called "tyrosemiophilia".

Fermentation Starter

A fermentation starter (called simply starter within the corresponding context, sometimes called a mother) is a preparation to assist the beginning of the fermentation process in preparation of various foods and fermented drinks. A starter culture is a microbiological culture which actually performs fermentation. These starters usually consist of a cultivation medium, such as grains, seeds, or nutrient liquids that have been well colonized by the microorganisms used for the fermentation.

Pain poolish - a type of fermentation starter for bread

In descriptions of national cuisines, fermentation starters may be referred to by their national names:

- Qū (simplified: 曲; traditional: 麴, also romanized as chu) (China)
    - Jiuqu (simplified Chinese: 酒曲; traditional Chinese: 酒麴; pinyin: *jiǔ qū*): the starter used for making Chinese alcoholic beverages
    - Laomian (pinyin: *laomian*; literally: "old dough" pinyin: *mianfei*; literally: "dough fat"): Chinese sourdough starter commonly used in Northern Chinese cuisine, the sourness of the starter is commonly quenched with sodium carbonate prior to use.
- Nuruk or Nulook (누룩), meju or Mae-joo or Mae-zu (메주) (Korea)
- Koji (ｒ) (Japan)
- Ragi (Indonesia and Malaysia)
- Bakhar, ranu, marchaar (murcha), Virjan (India)
- Bubod (Philippines)
- Luk paeng (Thai: ลูกแป้ง) (Thailand)
- Levain (France)
- Bread *zakvaska* (закваска, sourdough) (Russia, Ukraine)
- Opara (опара), (Russia), a starter based on yeast

These starters are formed using a specific cultivation medium and a specific mix of fungal and bacterial strains.

Typical microorganisms used in starters include various bacteria and fungi (yeasts and molds): *Rhizopus, Aspergillus, Mucor, Amylomyces, Endomycopsis, Saccharomyces, Hansenula anomala, Lactobacillus, Acetobacter*, etc. Various national cultures have various active ingredients in starters, and often involve mixed microflora.

Industrial starters include various enzymes, in addition to microflora.

Food Microbiology

Food microbiology is the study of the microorganisms that inhabit, create, or contaminate food, including the study of microorganisms causing food spoilage. "Good" bacteria, however, such as probiotics, are becoming increasingly important in food science. In addition, microorganisms are essential for the production of foods such as cheese, yogurt, bread, beer, wine and, other fermented foods.

Food Safety

Food safety is a major focus of food microbiology. Pathogenic bacteria, viruses and toxins produced by microorganisms are all possible contaminants of food. However, microorganisms and their products can also be used to combat these pathogenic microbes. Probiotic bacteria, including those that produce bacteriocins, can kill and inhibit pathogens. Alternatively, purified bacteriocins such as nisin can be added directly to food products. Finally, bacteriophages, viruses that only infect bacteria, can be used to kill bacterial pathogens. Thorough preparation of food, including proper cooking, eliminates most bacteria and viruses. However, *toxins produced* by contaminants may not be liable to change to non-toxic forms by heating or cooking the contaminated food due to other safety conditions.

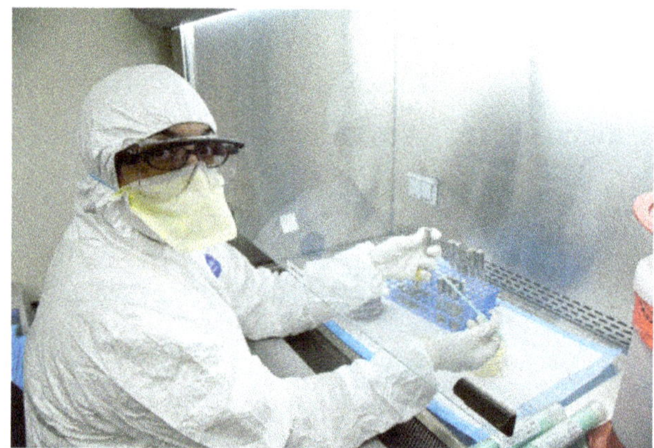

A microbiologist working in a biosafety laboratory tests for high risk pathogens in food.

Fermentation

Fermentation is one of the methods to preserve food and alter its quality. Yeast, especially *Saccharomyces cerevisiae*, is used to leaven bread, brew beer and make wine. Certain bacteria, including lactic acid bacteria, are used to make yogurt, cheese, hot sauce, pickles, fermented sausages and dishes such as kimchi. A common effect of these fermentations is that the food product is less hospitable to other microorganisms, including pathogens and spoilage-causing microorganisms, thus extending the food's shelf-life. Some cheese varieties also require molds to ripen and develop their characteristic flavors.

Microbial Biopolymers

Several microbially produced biopolymers are used in the food industry.

Alginate

Alginates can be used as thickening agents. Although listed here under the category 'Microbial polysaccharides', commercial alginates are currently only produced by extraction from brown seaweeds such as *Laminaria hyperborea* or *L. japonica*.

Poly-γ-glutamic Acid

Poly-γ-glutamic acid (γ-PGA) produced by various strains of *Bacillus* has potential applications as a thickener in the food industry.

Food Authenticity

It is important to be able to detect microorganisms in food, in particular pathogenic microorganisms or genetically modified microorganisms.

Food Testing

Food microbiology laboratory at the Faculty of Food Technology, Latvia University of Agriculture.

To ensure safety of food products, microbiological tests such as testing for pathogens and spoilage organisms are required. This way the risk of contamination under normal use conditions can be examined and food poisoning outbreaks can be prevented. Testing of food products and ingredients is important along the whole supply chain as possible flaws of products can occur at every stage of production. Apart from detecting spoilage, microbiological tests can also determine germ content, identify yeasts and molds, and salmonella. For salmonella, scientists are also developing rapid and portable technologies capable of identifying unique variants of Salmonella .

Polymerase Chain Reaction (PCR) is a quick and inexpensive method to generate numbers of copies of a DNA fragment at a specific band ("PCR (Polymerase Chain Reaction)," 2008). For that reason, scientists are using PCR to detect different kinds of viruses or bacteria, such as HIV and anthrax based on their unique DNA patterns. Various kits are commercially available to help in food pathogen nucleic acids extraction, PCR detection, and differentiation. The detection of bacterial strands in food products is very important to everyone in the world, for it helps prevent the occurrence of food borne illness. Therefore, PCR is recognized as a DNA detector in order to amplify and trace the presence of pathogenic strands in different processed food.

Brewing

Brewing is the production of beer by steeping a starch source (commonly cereal grains, the most popular of which is barley) in water and fermenting the resulting sweet liquid with yeast. It may be done in a brewery by a commercial brewer, at home by a homebrewer, or by a variety of traditional methods such as communally by the indigenous peoples in Brazil when making cauim. Brewing has taken place since around the 6th millennium BC, and archaeological evidence suggests that emerging civilizations including ancient Egypt and Mesopotamia brewed beer. Since the nineteenth century the brewing industry has been part of most western economies.

The basic ingredients of beer are water and a fermentable starch source such as malted barley. Most beer is fermented with a brewer's yeast and flavoured with hops. Less widely used starch sources include millet, sorghum and cassava. Secondary sources (adjuncts), such as maize (corn), rice, or sugar, may also be used, sometimes to reduce cost, or to add a feature, such as adding wheat to aid in retaining the foamy head of the beer. The proportion of each starch source in a beer recipe is collectively called the grain bill.

Steps in the brewing process include malting, milling, mashing, lautering, boiling, fermenting, conditioning, filtering, and packaging. There are three main fermentation methods, warm, cool and spontaneous. Fermentation may take place in an open or closed fermenting vessel; a secondary fermentation may also occur in the cask or bottle. There are several additional brewing methods, such as barrel aging, double dropping, and Yorkshire Square.

A 16th-century brewery

History

Brewing has taken place since around the 6th millennium BC, and archaeological evidence suggests emerging civilizations including ancient Egypt and Mesopotamia brewed beer. Descriptions of various beer recipes can be found in cuneiform (the oldest known writing) from ancient Mesopotamia. In Mesopotamia the brewer's craft was the only profession which derived social sanction and divine protection from female deities/goddesses, specifically: Ninkasi, who covered the production of beer, Siris, who was used in a metonymic way to refer to beer, and Siduri, who covered the enjoyment of beer.

As almost any cereal containing certain sugars can undergo spontaneous fermentation due to wild yeasts in the air, it is possible that beer-like beverages were independently developed throughout the world soon after a tribe or culture had domesticated cereal. Chemical tests of ancient pottery jars reveal that beer was produced as far back as about 7,000 years ago in what is today Iran. This discovery reveals one of the earliest known uses of fermentation and is the earliest evidence of brewing to date. In Mesopotamia, the oldest evidence of beer is believed to be a 6,000-year-old Sumerian tablet depicting people drinking a beverage through reed straws from a communal bowl. A 3900-year-old Sumerian poem honouring Ninkasi, the patron goddess of brewing, contains the oldest surviving beer recipe, describing the production of beer from barley via bread.

The invention of bread and beer has been argued to be responsible for humanity's ability to develop technology and build civilization. The earliest chemically confirmed barley beer to date was discovered at Godin Tepe in the central Zagros Mountains of Iran, where fragments of a jug, at least 5,000 years old was found to be coated with beerstone, a by-product of the brewing process.

Beer may have been known in Neolithic Europe as far back as 5,000 years ago, and was mainly brewed on a domestic scale.

Ale produced before the Industrial Revolution continued to be made and sold on a domestic scale, although by the 7th century AD beer was also being produced and sold by European monasteries. During the Industrial Revolution, the production of beer moved from artisanal manufacture to industrial manufacture, and domestic manufacture ceased to be significant by the end of the 19th century.The development of hydrometers and thermometers changed brewing by allowing the brewer more control of the process, and greater knowledge of the results.

Today, the brewing industry is a global business, consisting of several dominant multinational companies and many thousands of smaller producers ranging from brewpubs to regional breweries. More than 133 billion litres (35 billion gallons) are sold per year—producing total global revenues of $294.5 billion (£147.7 billion) in 2006.

Ingredients

Malted barley before kilning or roasting

The basic ingredients of beer are water; a starch source, such as malted barley, able to be fermented (converted into alcohol); a brewer's yeast to produce the fermentation; and a flavouring, such as hops, to offset the sweetness of the malt. A mixture of starch sources may be used, with a secondary saccharide, such as maize (corn), rice, or sugar, often being termed an adjunct, especially when used as a lower-cost substitute for malted barley. Less widely used starch sources include millet, sorghum, and cassava root in Africa, potato in Brazil, and agave in Mexico, among others. The amount of each starch source in a beer recipe is collectively called the grain bill.

Water

Beer is composed mostly of water. Regions have water with different mineral components; as a result, different regions were originally better suited to making certain types of beer, thus giving them a regional character. For example, Dublin has hard water well suited to making stout, such as Guinness; while Pilsen has soft water well suited to making pale lager, such as Pilsner Urquell. The waters of Burton in England contain gypsum, which benefits making pale ale to such a degree that brewers of pale ales will add gypsum to the local water in a process known as Burtonisation.

Starch Source

The starch source in a beer provides the fermentable material and is a key determinant of the strength and flavour of the beer. The most common starch source used in beer is malted grain. Grain is malted by soaking it in water, allowing it to begin germination, and then drying the partially germinated grain in a kiln. Malting grain produces enzymes that will allow conversion from starches in the grain into fermentable sugars during the mash process. Different roasting times and temperatures are used to produce different colours of malt from the same grain. Darker malts will produce darker beers.

Nearly all beer includes barley malt as the majority of the starch. This is because of its fibrous husk, which is important not only in the sparging stage of brewing (in which water is washed over the mashed barley grains to form the wort) but also as a rich source of amylase, a digestive enzyme that facilitates conversion of starch into sugars. Other malted and unmalted grains (including wheat, rice, oats, and rye, and, less frequently, maize (corn) and sorghum) may be used. In recent years, a few brewers have produced gluten-free beer made with sorghum with no barley malt for people who cannot digest gluten-containing grains like wheat, barley, and rye.

Hops

Hops are the female flower clusters or seed cones of the hop vine *Humulus lupulus*, which are used as a flavouring and preservative agent in nearly all beer made today. Hops had been used for medicinal and food flavouring purposes since Roman times; by the 7th century in Carolingian monasteries in what is now Germany, beer was being made with hops, though it isn't until the thirteenth century that widespread cultivation of hops for use in beer is recorded. Before the thirteenth century, beer was flavoured with plants such as yarrow, wild rosemary, and bog myrtle, and other ingredients such as juniper berries, aniseed and ginger, which would be combined into a mixture known as gruit and used as hops are now used; between the thirteenth and the sixteenth century, during which hops took over as the dominant flavouring, beer flavoured with gruit was known as ale, while beer flavoured with hops was known as beer. Some beers today, such as *Fraoch* by the Scottish Heather Ales company and *Cervoise Lancelot* by the French Brasserie-Lancelot company, use plants other than hops for flavouring.

Hop cone in a Hallertau, Germany, hop yard

Hops contain several characteristics that brewers desire in beer: they contribute a bitterness that balances the sweetness of the malt; they provide floral, citrus, and herbal aromas and flavours; they have an antibiotic effect that favours the activity of brewer's yeast over less desirable microorganisms; and they aid in "head retention", the length of time that a foamy head will last. The preservative in hops comes from the lupulin glands which contain soft resins with alpha and beta acids. Though much studied, the preservative nature of the soft resins is not yet fully understood, though it has been observed that unless stored at a cool temperature, the preservative nature will decrease. Brewing is the sole major commercial use of hops.

Yeast

Yeast is the microorganism that is responsible for fermentation in beer. Yeast metabolises the sugars extracted from grains, which produces alcohol and carbon dioxide, and thereby turns wort into beer. In addition to fermenting the beer, yeast influences the character and flavour. The dominant types of yeast used to make beer are *Saccharomyces cerevisiae*, known as ale yeast, and *Saccharomyces pastorianus*, known as lager yeast; *Brettanomyces* ferments lambics, and *Torulaspora delbrueckii* ferments Bavarian weissbier. Before the role of yeast in fermentation was understood, fermentation involved wild or airborne yeasts, and a few styles such as lambics still use this method today. Emil Christian Hansen, a Danish biochemist employed by the Carlsberg Laboratory, developed pure yeast cultures which were introduced into the Carlsberg brewery in 1883, and pure yeast strains are now the main fermenting source used worldwide.

Clarifying Agent

Some brewers add one or more clarifying agents to beer, which typically precipitate (collect as a solid) out of the beer along with protein solids and are found only in trace amounts in the finished product. This process makes the beer appear bright and clean, rather than the cloudy appearance of ethnic and older styles of beer such as wheat beers.

Examples of clarifying agents include isinglass, obtained from swimbladders of fish; Irish moss, a seaweed; kappa carrageenan, from the seaweed *Kappaphycus cottonii*; Polyclar (artificial); and gelatin. If a beer is marked "suitable for Vegans", it was generally clarified either with seaweed or with artificial agents, although the "Fast Cask" method invented by Marston's in 2009 may provide another method.

Brewing Process

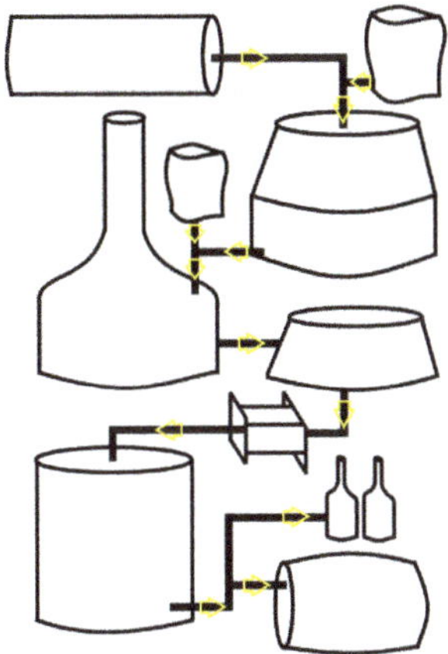

Diagram illustrating the process of brewing beer
Hot Water Tank
Mash Tun
Malt
Hops
Copper
Hopback
Add Yeast toFermenter
Heat
exchanger
Bottling
Cask or Keg

There are several steps in the brewing process, which may include malting, mashing, lautering, boiling, fermenting, conditioning, filtering, and packaging.

Malting is the process where barley grain is made ready for brewing. Malting is broken down into three steps in order to help to release the starches in the barley. First, during steeping, the grain is added to a vat with water and allowed to soak for approximately 40 hours. During germination, the grain is spread out on the floor of the germination room for around 5 days. The final part of malting is kilning when the malt goes through a very high temperature drying in a kiln; with gradual temperature increase over several hours.

When kilning is complete, the grains are now termed malt, and they will be milled or crushed to break apart the kernels and expose the cotyledon, which contains the majority of the carbohydrates and sugars; this makes it easier to extract the sugars during mashing. Milling also separates the seed from the husk. Care must be taken when milling to ensure that the starch reserves are sufficiently milled without damaging the husk and providing coarse enough grits that a good filter bed can be formed during lautering. Grains are typically dry-milled with roller mills or hammer mills. Hammer mills, which produce a very fine mash, are often used when mash filters are going to be employed in the lautering process because the grain does not have to form its own filter bed. In modern plants, the grain is often conditioned with water before it is milled to make the husk more pliable, thus reducing breakage and improving lauter speed.

Mashing converts the starches released during the malting stage into sugars that can be fermented. The milled grain is mixed with hot water in a large vessel known as a mash tun. In this vessel, the grain and water are mixed together to create a cereal mash. During the mash, naturally occurring enzymes present in the malt convert the starches (long chain carbohydrates) in the grain into smaller molecules or simple sugars (mono- , di-, and tri-saccharides). This "conversion" is called saccharification. The result of the mashing process is a sugar-rich liquid or "wort", which is then strained through the bottom of the mash tun in a process known as lautering. Prior to lautering, the mash temperature may be raised to about 75–78 °C (167–172 °F) (known as a mashout) to free up more starch and reduce mash viscosity. Additional water may be sprinkled on the grains to extract additional sugars (a process known as sparging).

The wort is moved into a large tank known as a "copper" or kettle where it is boiled with hops and sometimes other ingredients such as herbs or sugars. This stage is where many chemical and technical reactions take place, and where important decisions about the flavour, colour, and aroma of the beer are made. The boiling process serves to terminate enzymatic processes, precipitate proteins, isomerize hop resins, and concentrate and sterilize the wort. Hops add flavour, aroma and bitterness to the beer. At the end of the boil, the hopped wort settles to clarify in a vessel called a "whirlpool", where the more solid particles in the wort are separated out.

After the whirlpool, the wort is drawn away from the compacted hop trub, and rapidly cooled via a heat exchanger to a temperature where yeast can be added. A variety of heat exchanger designs are used in breweries, with the most common a plate-style. Water or glycol run in channels in the opposite direction of the wort, causing a rapid drop in temperature. It is very important to quickly cool the wort to a level where yeast can be added safely as yeast is unable to grow in very high temperatures, and will start to die in temperatures above 60°C. After the wort goes through the heat exchanger, the cooled wort goes into a fermentation tank. A type of yeast is selected and added, or "pitched", to the fermentation tank. When the yeast is added to the wort, the fermenting process begins, where the sugars turn into alcohol, carbon dioxide and other components. When the fermentation is complete the brewer may rack the beer into a new tank, called a conditioning tank. Conditioning of the beer is the process in which the beer ages, the flavour becomes smoother, and flavours that are unwanted dissipate. After conditioning for a week to several months, the beer may be filtered and force carbonated for bottling, or fined in the cask.

Mashing

A mash tun at the Bass Museum in Burton-upon-Trent

Mashing is the process of combining a mix of milled grain (typically malted barley with supplementary grains such as corn, sorghum, rye or wheat), known as the "grain bill", and water, known as "liquor", and heating this mixture in a vessel called a "mash tun". Mashing is a form of steeping, and defines the act of brewing, such as with making tea, sake, and soy sauce. Technically, wine, cider and mead are not brewed but rather vinified, as there is no steeping process involving solids. Mashing allows the enzymes in the malt to break down the starch in the grain into sugars, typically maltose to create a malty liquid called wort. There are two main methods – infusion mashing, in which the grains are heated in one vessel; and decoction mashing, in which a proportion of the grains are boiled and then returned to the mash, raising the temperature. Mashing involves pauses at certain temperatures (notably 45–62–73 °C or 113–144–163 °F), and takes place in a "mash tun" – an insulated brewing vessel with a false bottom. The end product of mashing is called a "mash".

Mashing usually takes 1 to 2 hours, and during this time the various temperature rests activate different enzymes depending upon the type of malt being used, its modification level, and the intention of the brewer. The activity of these enzymes convert the starches of the grains to dextrins and then to fermentable sugars such as maltose. A mash rest from 49–55 °C (120–131 °F) activates various proteases, which break down proteins that might otherwise cause the beer to be hazy. This rest is generally used only with undermodified (i.e. undermalted) malts which are decreasingly popular in Germany and the Czech Republic, or non-malted grains such as corn and rice, which are widely used in North American beers. A mash rest at 60 °C (140 °F) activates β-glucanase, which breaks down gummy β-glucans in the mash, making the sugars flow out more freely later in the process. In the modern mashing process, commercial fungal based β-glucanase may be added as a supplement. Finally, a mash rest temperature of 65–71 °C (149–160 °F) is used to convert the starches in the malt to sugar, which is then usable by the yeast later in the brewing process. Doing the latter rest at the lower end of the range favours β-amylase enzymes, producing more low-order sugars like maltotriose, maltose, and glucose which are more fermentable by the yeast. This in turn creates a beer lower in body and higher in alcohol. A rest closer to the higher end of the range favours α-amylase enzymes, creating more higher-order sugars and dextrins which are less fermentable by the yeast, so a fuller-bodied beer with less alcohol is the result. Duration and pH variances also affect the sugar composition of the resulting wort.

Lautering

Lautering is the separation of the wort (the liquid containing the sugar extracted during mashing) from the grains. This is done either in a mash tun outfitted with a false bottom, in a lauter tun, or in a mash filter. Most separation processes have two stages: first wort run-off, during which the extract is separated in an undiluted state from the spent grains, and sparging, in which extract which remains with the grains is rinsed off with hot water. The lauter tun is a tank with holes in the bottom small enough to hold back the large bits of grist and hulls. The bed of grist that settles on it is the actual filter. Some lauter tuns have provision for rotating rakes or knives to cut into the bed of grist to maintain good flow. The knives can be turned so they push the grain, a feature used to drive the spent grain out of the vessel. The mash filter is a plate-and-frame filter. The empty frames contain the mash, including the spent grains, and have a capacity of around one hectoliter. The plates contain a support structure for the filter cloth. The plates, frames, and filter cloths are arranged in a carrier frame like so: frame, cloth, plate, cloth, with plates at each end of the structure. Newer mash filters have bladders that can press the liquid out of the grains between spargings. The grain does not act like a filtration medium in a mash filter.

Boiling

After mashing, the beer wort is boiled with hops (and other flavourings if used) in a large tank known as a "copper" or brew kettle – though historically the mash vessel was used and is still in some small breweries. The boiling process is where chemical and technical reactions take place, including sterilization of the wort to remove unwanted bacteria, releasing of hop flavours, bitterness and aroma compounds through isomerization, stopping of enzymatic processes, precipitation of proteins, and concentration of the wort. Finally, the vapours produced during the boil volatilise off-flavours, including dimethyl sulfide precursors. The boil is conducted so that it is even and intense – a continuous "rolling boil". The boil on average lasts between 45 and 90 minutes, depending on its intensity, the hop addition schedule, and volume of water the brewer expects to evaporate. At the end of the boil, solid particles in the hopped wort are separated out, usually in a vessel called a "whirlpool".

Brew Kettle or Copper

Copper is the traditional material for the boiling vessel, because copper transfers heat quickly and evenly, and because the bubbles produced during boiling, and which would act as an insulator against the heat, do not cling to the surface of copper, so the wort is heated in a consistent manner. The simplest boil kettles are direct-fired, with a burner underneath. These can produce a vigorous and favourable boil, but are also apt to scorch the wort where the flame touches the kettle, causing caramelisation and making cleanup difficult. Most breweries use a steam-fired kettle, which uses steam jackets in the kettle to boil the wort. Breweries usually have a boiling unit either inside or outside of the kettle, usually a tall, thin cylinder with vertical tubes, called a calandria, through which wort is pumped.

Brew kettles at Coors Brewing Company

Whirlpool

At the end of the boil, solid particles in the hopped wort are separated out, usually in a vessel called a "whirlpool" or "settling tank". The whirlpool was devised by the Molson Brewery in 1960 to utilise the so-called tea leaf paradox to force the denser solids known as "trub" (coagulated proteins, vegetable matter from hops) into a cone in the centre of the whirlpool tank. Whirlpool systems vary: smaller breweries tend to use the brew kettle, larger breweries use a separate tank, and design will differ, with tank floors either flat, sloped, conical or with a cup in the centre. The principle in all is that by swirling the wort the centripetal force will push the trub into a cone at the centre of the bottom of the tank, where it can be easily removed.

Hopback

A hopback is a traditional additional chamber that acts as a sieve or filter by using whole hops to clear debris (or "trub") from the unfermented (or "green") wort, as the whirlpool does, and also to increase hop aroma in the finished beer. It is a chamber between the brewing kettle and wort chiller. Hops are added to the chamber, the hot wort from the kettle is run through it, and then immediately cooled in the wort chiller before entering the fermentation chamber. Hopbacks utilizing a sealed chamber facilitate maximum retention of volatile hop aroma compounds that would normally be driven off when the hops contact the hot wort. While a hopback has a similar filtering effect as a whirlpool, it operates differently: a whirlpool uses centrifugal forces, a hopback uses a layer of whole hops to act as a filter bed. Furthermore, while a whirlpool is useful only for the removal of pelleted hops (as flowers do not tend to separate as easily), in general hopbacks are used only for the removal of whole flower hops (as the particles left by pellets tend to make it through the hopback). The hopback has mainly been substituted in modern breweries by the whirlpool.

Wort Cooling

After the whirlpool, the wort must be brought down to fermentation temperatures (20–26 °C) before yeast is added. In modern breweries this is achieved through a plate heat exchanger. A plate heat exchanger has many ridged plates, which form two separate paths. The wort is pumped into the heat exchanger, and goes through every other gap between the plates. The cooling medium, usually water, goes through the other gaps. The ridges in the plates ensure turbulent flow. A good heat exchanger can drop 95 °C wort to 20 °C while warming the cooling medium from about 10 °C to 80 °C. The last few plates often use a cooling medium which can be cooled to below the freezing point, which allows a finer control over the wort-out temperature, and also enables cooling to around 10 °C. After cooling, oxygen is often dissolved into the wort to revitalize the yeast and aid its reproduction.

While boiling, it is useful to recover some of the energy used to boil the wort. On its way out of the brewery, the steam created during the boil is passed over a coil through which unheated water flows. By adjusting the rate of flow, the output temperature of the water can be controlled. This is also often done using a plate heat exchanger. The water is then stored for later use in the next mash, in equipment cleaning, or wherever necessary. Another common method of energy recovery takes place during the wort cooling. When cold water is used to cool the wort in a heat exchanger, the water is significantly warmed. In an efficient brewery, cold water is passed through the heat exchanger at a rate set to maximize the water's temperature upon exiting. This now-hot water is then stored in a hot water tank.

Fermenting

Fermentation takes place in fermentation vessels which come in various forms, from enormous cylindroconical vessels, through open stone vessels, to wooden vats. After the wort is cooled and aerated – usually with sterile air – yeast is added to it, and it begins to ferment. It is during this stage that sugars won from the malt are converted into alcohol and carbon dioxide, and the product can be called beer for the first time.

Most breweries today use cylindroconical vessels, or CCVs, which have a conical bottom and a cylindrical top. The cone's aperture is typically around 60°, an angle that will allow the yeast to flow towards the cone's apex, but is not so steep as to take up too much vertical space. CCVs can handle both fermenting and conditioning in the same tank. At the end of fermentation, the yeast and other solids which have fallen to the cone's apex can be simply flushed out of a port at the apex. Open fermentation vessels are also used, often for show in brewpubs, and in Europe in wheat beer fermentation. These vessels have no tops, which makes harvesting top-fermenting yeasts very easy. The open tops of the vessels make the risk of infection greater, but with proper cleaning procedures and careful protocol about who enters fermentation chambers, the risk can be well controlled. Fermentation tanks are typically made of stainless steel. If they are simple cylindrical tanks with beveled ends, they are arranged vertically, as opposed to conditioning tanks which are usually laid out horizontally. Only a very few breweries still use wooden vats for fermentation as wood is difficult to keep clean and infection-free and must be repitched more or less yearly.

Modern closed fermentation vessels

Fermentation Methods

Open vessel showing fermentation taking place

There are three main fermentation methods, warm, cool and wild or spontaneous. Fermentation may take place in open or closed vessels. There may be a secondary fermentation which can take place in the brewery, in the cask or in the bottle.

Brewing yeasts are traditionally classed as "top-cropping" (or "top-fermenting") and "bottom-cropping" (or "bottom-fermenting"). Yeast were termed top or bottom cropping, because in traditional brewing yeast was collected from the top or bottom of the fermenting wort to be reused for the next brew. This terminology is somewhat inappropriate in the modern era; after the widespread application of brewing mycology it was discovered that the two separate collecting methods involved two different yeast species that favoured different temperature regimes, namely *Saccharomyces cerevisiae* in top-cropping at warmer temperatures and *Saccharomyces pastorianus* in bottom-cropping at cooler temperatures. As brewing methods changed in the 20th century, cylindro-conical fermenting vessels became the norm and the collection of yeast for both *Saccharomyces* species is done from the bottom of the fermenter, thus the method of collection no longer implies a species association. There are a few remaining breweries who collect yeast in the top-cropping method, such as Samuel Smiths brewery in Yorkshire, Marstons in Staffordshire and several German hefeweizen producers.

For both types, yeast is fully distributed through the beer while it is fermenting, and both equally flocculate (clump together and precipitate to the bottom of the vessel) when fermentation is finished. By no means do all top-cropping yeasts demonstrate this behaviour, but it features strongly in many English yeasts that may also exhibit chain forming (the failure of budded cells to break from the mother cell), which is in the technical sense different from true flocculation. The most common top-cropping brewer's yeast, *Saccharomyces cerevisiae*, is the same species as the common baking yeast. However, baking and brewing yeasts typically belong to different strains, cultivated to favour different characteristics: baking yeast strains are more aggressive, in order to carbonate dough in the shortest amount of time; brewing yeast strains act slower, but tend to tolerate higher alcohol concentrations (normally 12–15% abv is the maximum, though under special treatment some ethanol-tolerant strains can be coaxed up to around 20%). Modern quantitative genomics has revealed the complexity of *Saccharomyces* species to the extent that yeasts involved in beer and wine production commonly involve hybrids of so-called pure species. As such, the yeasts involved in what has been typically called top-cropping or top-fermenting ale may be both *Saccharomyces cerevisiae* and complex hybrids of *Saccharomyces cerevisiae* and *Saccharomyces kudriavzevii*. Three notable ales, Chimay, Orval and Westmalle, are fermented with these hybrid strains, which are identical to wine yeasts from Switzerland.

Warm Fermentation

In general, yeasts such as *Saccharomyces cerevisiae* are fermented at warm temperatures between 15 and 20 °C (59 and 68 °F), occasionally as high as 24 °C (75 °F), while the yeast used by Brasserie Dupont for saison ferments even higher at 29 to 35 °C (84 to 95 °F). They generally form a foam on the surface of the fermenting beer, as during the fermentation process its hydrophobic surface causes the flocs to adhere to CO_2 and rise; because of this, they are often referred to as "top-cropping" or "top-fermenting" – though this distinction is less clear in modern brewing with the use of cylindro-conical tanks. Generally, warm-fermented beers are ready to drink within three weeks after the beginning of fermentation, although some brewers will condition them for several months.

Cool Fermentation

Lager is beer that has been cool fermented at around 10 °C (50 °F) (compared to typical warm fermentation temperatures of 18 °C (64 °F)), then stored for around 30 days at temperatures close to freezing point; during this storage sulphur components developed during fermentation dissipate. Though it is the cool fermentation that defines lager, the main technical difference with lager yeast, *Saccharomyces pastorianus*, is its divergent genome and its ability to metabolize both melibiose, a disaccharide of galactose and glucose and raffinose (a trisaccharide composed of the sugars galactose, fructose, and glucose).Ale yeasts, *Saccharomyces cerevisiae*, can only partially metabolize raffinose and cannot metabolize melibiose at all. Nonetheless, these sugars are not present in typical beer wort made from malted barley and their metabolism or lack of it will not affect the subsequent beer organoleptic qualities in any way.

Brewers in Bavaria had for centuries been selecting cold-fermenting yeasts by storing ("lagern") their beers in cold alpine caves. The process of natural selection meant that the wild yeasts that were most cold tolerant would be the ones that would remain actively fermenting in the beer that was stored in the caves. A sample of these Bavarian yeasts was sent from the Spaten brewery in Munich to the Carlsberg brewery in Copenhagen in 1845 who began brewing with it. In 1883 Emile Hansen completed a study on pure yeast culture isolation and the pure strain obtained from Spaten went into industrial production in 1884 as Carlsberg yeast No 1. Another specialized pure yeast production plant was soon installed at the Heineken Brewery in Rotterdam the following year and together they began supply of pure cultured yeast to brewers across Europe. This yeast strain was originally classified as *Saccharomyces carlsbergensis*, a now defunct species name which has been superseded by the currently accepted taxonomic classification *Saccharomyces pastorianus*.

Today, lagers represent the vast majority of beers produced. Examples include Budweiser Budvar, Birra Moretti, Stella Artois, Red Stripe, and Singha. Some lagers are marketed as Pilsner, which originated in Pilsen, Czech Republic *(Plzeň in Czech)*.

Lager yeast normally ferments at a temperature of approximately 5 °C (41 °F). Lager yeast can be fermented at a higher temperature normally used for top-fermenting yeast, and this application is often used in a beer style known as *California Common* or colloquially as "steam beer".

Spontaneous Fermentation

Lambic beers are historically brewed in Brussels and the nearby Pajottenland region of Belgium without any yeast inoculation. They are fermented in oak barrels with the resident microbiota present in the wood and can take up to 2 years to come into condition for sale. However, with the advent of yeast banks and England's National Collection of Yeast Cultures, brewing these beers – albeit not through spontaneous fermentation –

is possible anywhere. Specific bacteria cultures are also available to reproduce certain styles. *Brettanomyces* is a genus of yeast important in brewing lambic, a beer produced not by the deliberate addition of brewer's yeasts, but by spontaneous fermentation with wild yeasts and bacteria.

Conditioning

Conditioning tanks at Anchor Brewing Company

After an initial or primary fermentation, beer is *conditioned*, matured or aged, in one of several ways, which can take from 2 to 4 weeks, several months, or several years, depending on the brewer's intention for the beer. The beer is usually transferred into a second container, so that it is no longer exposed to the dead yeast and other debris (also known as "trub") that have settled to the bottom of the primary fermenter. This prevents the formation of unwanted flavours and harmful compounds such as acetylaldehyde.

Kräusening

Kräusening is a conditioning method in which fermenting wort is added to the finished beer. The active yeast will restart fermentation in the finished beer, and so introduce fresh carbon dioxide; the conditioning tank will be then sealed so that the carbon dioxide is dissolved into the beer producing a lively "condition" or level of carbonation. The kräusening method may also be used to condition bottled beer.

Lagering

Lagers are stored at near freezing temperatures for 1–6 months while still on the yeast. The process of storing, or conditioning, or maturing, or aging a beer at a low temperature for a long period is called "lagering", and while it is associated with lagers, the process may also be done with ales, with the same result – that of cleaning up various chemicals, acids and compounds.

Secondary fermentation

During secondary fermentation, most of the remaining yeast will settle to the bottom of the second fermenter, yielding a less hazy product.

Bottle fermentation

Some beers undergo a fermentation in the bottle, giving natural carbonation. This may be a second or third fermentation. They are bottled with a viable yeast population in suspension. If there is no residual fermentable sugar left, sugar or wort or both may be added in a process known as priming. The resulting fermentation generates CO_2 that is trapped in the bottle, remaining in solution and providing natural carbonation. Bottle-conditioned beers may be either filled unfiltered direct from the fermentation or conditioning tank, or filtered and then reseeded with yeast.

Cask conditioning

Cask ale or cask-conditioned beer is unfiltered and unpasteurised beer that is conditioned (including secondary fermentation) and served from a cask, either pumped up from a cellar via a beer engine (hand pump), or from a tap by gravity. Sometimes a cask breather is used to keep the beer fresh by allowing carbon dioxide to replace oxygen as the beer is drawn off the cask. The term "real ale" as used by the Campaign for Real Ale (CAMRA) refers to beer "served without the use of extraneous carbon dioxide", which would disallow the use of a cask breather.

Filtering

A mixture of diatomaceous earth and yeast after filtering

Filtering the beer stabilizes the flavour, and gives beer its polished shine and brilliance. Not all beer is filtered. When tax determination is required by local laws, it is typically done at this stage in a calibrated tank. There are several forms of filters, they may be in the form of sheets or "candles", or they may be a fine powder such as diatomaceous earth, also called kieselguhr. The powder is added to the beer and recirculated past screens to form a filtration bed.

Filters range from rough filters that remove much of the yeast and any solids (e.g., hops, grain particles) left in the beer, to filters tight enough to strain colour and body from the beer. Filtration ratings are divided into rough, fine, and sterile. Rough filtration leaves some cloudiness in the beer, but it is noticeably clearer than unfiltered beer. Fine filtration removes almost all cloudiness. Sterile filtration removes almost all microorganisms.

Sheet (pad) filters

These filters use sheets that allow only particles smaller than a given size to pass through. The sheets are placed into a filtering frame, sanitized (with boiling water, for example) and then used to filter the beer. The sheets can be flushed if the filter becomes blocked. The sheets are usually disposable and are replaced between filtration sessions. Often the sheets contain powdered filtration media to aid in filtration.

Pre-made filters have two sides. One with loose holes, and the other with tight holes. Flow goes from the side with loose holes to the side with the tight holes, with the intent that large particles get stuck in the large holes while leaving enough room around the particles and filter medium for smaller particles to go through and get stuck in tighter holes.

Sheets are sold in nominal ratings, and typically 90% of particles larger than the nominal rating are caught by the sheet.

Kieselguhr filters

Filters that use a powder medium are considerably more complicated to operate, but can filter much more beer before regeneration. Common media include diatomaceous earth and perlite.

Packaging

Packaging is putting the beer into the containers in which it will leave the brewery. Typically, this means putting the beer into bottles, aluminium cans, kegs, or casks, but it may include putting the beer into bulk tanks for high-volume customers.

By-products

Brewing by-products are "spent grain" and the sediment (or "dregs") from the filtration process which may be dried and resold as "brewers dried yeast" for poultry feed, or made into yeast extract.

Yeast extract

Yeast extract is used in brands such as Vegemite and Marmite. The process of turning the yeast sediment into edible yeast extract was discovered by a German scientist Justus Liebig.

Spent grain

Brewer's spent grain (also called spent grain, brewer's grain or draff) is the main by-product of the brewing process, it consists of the residue of malt and grain which remains in the mash-kettle after the mashing and lautering process. It consists primarily of grain husks, pericarp, and fragments of endosperm. As it mainly consists of carbohydrates and proteins, and is readily consumed by animals, spent grain is used in animal feed. Spent grains can also be used as fertilizer, whole grains in bread, as well as in the production of biogas. Spent grain is also an ideal medium for growing mushrooms, such as shiitake, and already some breweries are either growing their own mushrooms or supplying spent grain to mushroom farms. Spent grains can be used in the production of red bricks, to improve the open porosity and reduce thermal conductivity of the ceramic mass.

Spent grain

Brewing Industry

The brewing industry is a global business, consisting of several dominant multinational companies and many thousands of smaller producers known as microbreweries or regional breweries depending on size and region. More than 133 billion liters (3.5×10^{10} U.S. gallons; 2.9×10^{10} imperial gallons) are sold per year—producing total global revenues of $294.5 billion (£147.7 billion) as of 2006. SABMiller became the largest brewing company in the world when it acquired Royal Grolsch, brewer of Dutch premium beer brand Grolsch. InBev was the second-largest beer-producing company in the world and Anheuser-Busch held the third spot, but after the acquisition of Anheuser-Busch by InBev, the new Anheuser-Busch InBev company is currently the largest brewer in the world.

Brewing at home is subject to regulation and prohibition in many countries. Restrictions on homebrewing were lifted in the UK in 1963, Australia followed suit in 1972, and the USA in 1978, though individual states were allowed to pass their own laws limiting production.

Fermentation in Winemaking

The process of fermentation in winemaking turns grape juice into an alcoholic beverage. During fermentation, yeasts transform sugars present in the juice into ethanol and carbon dioxide (as a by-product). In winemaking, the temperature and speed of fermentation are important considerations as well as the levels of oxygen present in the must at the start of the fermentation. The risk of stuck fermentation and the development of several wine faults can also occur during this stage, which can last anywhere from 5 to 14 days for *primary fermentation* and potentially another 5 to 10 days for a *secondary fermentation*. Fermentation may be done in stainless steel tanks, which is common with many white wines like Riesling, in an open wooden vat, inside a wine barrel and inside the wine bottle itself as in the production of many sparkling wines.

Fermenting must

History

The natural occurrence of fermentation means it was probably first observed long ago by humans. The earliest uses of the word "fermentation" in relation to winemaking was in reference to the apparent "boiling" within the must that came from the anaerobic reaction of the yeast to the sugars in the grape juice and the release of carbon dioxide. The Latin *fervere* means, literally, *to boil.* In the mid-19th century, Louis Pasteur noted the connection between yeast and the process of the fermentation in which the yeast act as catalyst and mediator through a series of a reaction that convert sugar into alcohol. The discovery of the Embden–Meyerhof–Parnas pathway by Gustav Embden, Otto Fritz Meyerhof and Jakub Karol Parnas in the early 20th century contributed more to the understanding of the complex chemical processes involved in the conversion of sugar to alcohol.

Process

In winemaking, there are distinctions made between *ambient yeasts* which are naturally present in wine cellars, vineyards and on the grapes themselves (sometimes known as a grape's "bloom" or "blush") and *cultured yeast* which are specifically isolated and inoculated for use in winemaking. The most common genera of wild yeasts found in winemaking include *Candida, Klöckera/Hanseniaspora, Metschnikowiaceae, Pichia* and *Zygosaccharomyces*. Wild yeasts can produce high-quality, unique-flavored wines; however, they are often unpredictable and may introduce less desirable traits to the wine, and can even contribute to spoilage. It should be noted that few yeast, and lactic and acetic acid bacterial colonies naturally live on the surface of grapes, but traditional wine makers, particularly in Europe, advocate use of ambient yeast as a characteristic of the region's *terroir*; nevertheless, many winemakers prefer to control fermentation with predictable cultured yeast. The cultured yeasts most commonly used in winemaking belong to the *Saccharomyces cerevisiae* (also known as "sugar yeast") species. Within this species are several hundred different strains of yeast that can be used during fermentation to affect the heat or vigor of the process and enhance or suppress certain flavor characteristics of the varietal. The use of different strains of yeasts is a major contributor to the diversity of wine, even among the same grape variety. Alternative, non-*Saccharomyces cerevisiae*, yeasts are being used more prevalently in the industry to add greater complexity to wine. After a winery has been in operation for a number of years, few yeast strains are actively involved in the fermentation process. The use of active dry yeasts reduces the variety of strains that appear in spontaneous fermentation by outcompeting those strains that are naturally present.

"Bloom", visible as a dusting on the berries

The addition of cultured yeast normally occurs with the yeast first in a dried or "inactive" state and is reactivated in warm water or diluted grape juice prior to being added to the must. To thrive and be active in fermentation, the yeast needs access to a continuous supply of carbon, nitrogen, sulfur, phosphorus as well as access to various vitamins and minerals. These components are naturally present in the grape must but their amount may be corrected by adding nutrients to the wine, in order to foster a more encouraging environment for the yeast. Newly formulated time-release nutrients, specifically manufactured for wine fermentations, offer the most advantageous conditions for yeast. Oxygen is needed as well, but in wine making, the risk of oxidation and the lack of alcohol production from oxygenated yeast requires the exposure of oxygen to be kept at a minimum.

Dry winemaking yeast (left) and yeast nutrients used in the rehydration process to stimulate yeast cells.

Upon the introduction of active yeasts to the grape must, phosphates are attached to the sugar and the six-carbon sugar molecules begin to be split into three-carbon pieces and go through a series of rearrangement reactions. During this process, the carboxylic carbon atom is released in the form of carbon dioxide with the remaining components becoming acetaldehyde. The absence of oxygen in this anaerobic process allows the acetaldehyde to be eventually converted, by reduction, to ethanol. During the conversion of acetaldehyde, a small amount is converted, by oxidation, to acetic acid which, in excess, can contribute to the wine fault known as volatile acidity (vinegar taint). After the yeast has exhausted its life cycle, they fall to the bottom of the fermentation tank as sediment known as lees. Yeast ceases its activity whenever all of the sugar in must has been converted into other chemicals or whenever the alcohol content has reached 15% alcohol per unit volume; a concentration strong enough to halt the enzymatic activity of almost all strains of yeast.

Other Compounds Involved

The metabolism of amino acids and breakdown of sugars by yeasts has the effect of creating other biochemical compounds that can contribute to the flavor and aroma of wine. These compounds can be considered "volatile" like aldehydes, ethyl acetate, ester, fatty acids, fusel oils, hydrogen sulfide, ketones and mercaptans or "non-volatile" like glycerol, acetic acid and succinic acid. Yeast also has the effect during fermentation of releasing glycoside hydrolase which can hydrolyse the flavor precursors of aliphatics (a flavor component that reacts with oak), benzene derivatives, monoterpenes (responsible for floral aromas from grapes like Muscat and Traminer), norisoprenoids (responsible for some of the spice notes in Chardonnay), and phenols.

Some strains of yeasts can generate volatile thiols which contribute to the fruity aromas in many wines such as the gooseberry scent commonly associated with Sauvignon blanc. *Brettanomyces* yeasts are responsible for the "barnyard aroma" characteristic in some red wines like Burgundy and Pinot noir.

Methanol is not a major constituent of wine. The usual concentration range is between 0.1 g/liter and 0.2 g/liter. These small traces have no adverse effect on people and no direct effect on the senses.

Winemaking Considerations

Carbon dioxide is visible during the fermentation process in the form of bubbles in the must.

During fermentation, there are several factors that winemakers take into consideration, with the most influential to ethanol production being sugar content in the must, the yeast strain used, and the fermentation temperature. The biochemical process of fermentation itself creates a lot of residual heat which can take the must out of the ideal temperature range for the wine. Typically, white wine is fermented between 18-20 °C (64-68 °F) though a wine maker may choose to use a higher temperature to bring out some of the complexity of the wine. Red wine is typically fermented at higher temperatures up to 29 °C (85 °F). Fermentation at higher temperatures may have adverse effect on the wine in stunning the yeast to inactivity and even "boiling off" some of the flavors of the wines. Some winemakers may ferment their red wines at cooler temperatures, more typical of white wines, in order to bring out more fruit flavors.

To control the heat generated during fermentation, the winemaker must choose a suitable vessel size or else use a cooling device. Various kinds of cooling devices are available, ranging from the ancient Bordeaux practice of placing the fermentation vat atop blocks of ice to sophisticated fermentation tanks that have built-in cooling rings.

A risk factor involved with fermentation is the development of chemical residue and spoilage which can be corrected with the addition of sulfur dioxide (SO_2), although excess SO_2 can lead to a wine fault. A winemaker who wishes to make a wine with high levels of residual sugar (like a dessert wine) may stop fermentation early either by dropping the temperature of the must to stun the yeast or by adding a high level of alcohol (like brandy) to the must to kill off the yeast and create a fortified wine.

The ethanol produced through fermentation acts as an important co-solvent to the non-polar compound that water cannot dissolve, such as pigments from grape skins, giving wine varieties their distinct color, and other aromatics. Ethanol and the acidity of wine act as an inhibitor to bacterial growth, allowing wine to be safely kept for years in the absence of air.

Other Types of Fermentation

A California Chardonnay that shows it has been barrel fermented.

In winemaking, there are different processes that fall under the title of "Fermentation" but might not follow the same procedure commonly associated with wine fermentation.

Bottle Fermentation

Bottle fermentation is a method of sparkling wine production, originating in the Champagne region where after the cuvee has gone through a primary yeast fermentation the wine is then bottled and goes through a secondary fermentation where sugar and additional yeast known as *liqueur de tirage* is added to the wine. This secondary fermentation is what creates the carbon dioxide bubbles that sparkling wine is known for.

Carbonic Maceration

The process of carbonic maceration is also known as whole grape fermentation where instead of yeast being added, the grapes fermentation is encouraged to take place inside the individual grape berries. This method is common in the creation of Beaujolais wine and involves whole clusters of grapes being stored in a closed container with the oxygen in the container being replaced with carbon dioxide. Unlike normal fermentation where yeast converts sugar into alcohol, carbonic maceration works by enzymes within the grape breaking down the cellular matter to form ethanol and other chemical properties. The resulting wines are typically soft and fruity.

Malolactic Fermentation

Instead of yeast, bacteria play a fundamental role in malolactic fermentation which is essentially the conversion of malic acid into lactic acid. This has the benefit of reducing some of the tartness and making the resulting wine taste softer. Depending on the style of wine that the winemaker is trying to produce, malolactic fermentation may take place at the very same time as the yeast fermentation. Alternatively, some strains of yeast may be developed that can convert L-malate to L-lactate during alcohol fermentation. For example, *Saccharomyces cerevisiae* strain ML01 (*S. cerevisiae strain* ML01), which carries a gene encoding malolactic enzyme from *Oenococcus oeni* and a gene encoding malate permease from *Schizosaccharomyces pombe. S. cerevisiae strain* ML01 has received regulatory approval in both Canada and the United States.

Fermentation Lock

The fermentation lock or airlock is a device used in beer brewing and wine making that allows carbon dioxide released during fermentation to escape the fermenter, while not allowing air to enter the fermenter, thus avoiding oxidation.

Fermentation lock on homebrewing fermentation vessel

There are two main designs for the fermentation lock, or airlock. These designs work when half filled with water. When the pressure of the gas inside the fermentation vessel exceeds the prevailing atmospheric pressure the gas will push its way through the water as individual bubbles into the outside air. A sanitizing solution or vodka is sometimes placed in the fermentation lock to prevent contamination of the beer in case the water is inadvertently drawn into the fermenter.

This device may take the form of a tube connected to the headspace of the fermenting vessel into a tub of sanitized liquid or a simpler device mounted directly on top of the fermentation vessel.

Currently, a popular fermentation lock that mounts on top of the fermentation vessel is the three-piece fermentation lock. Other models contain three bulbous chambers allowing for a broader range of pressure equalization. These bulbous fermentation locks were generally made of hand blown glass and are nowadays often made of clear plastic.

The use of perforated rubber balloons offers an easy and inexpensive alternative to conventional airlocks: as used primarily in homebrewing, the balloon is stretched over the orifice of the fermentation vessel and, if necessary, tightened with rubber bands. The balloon is then perforated with a needle. These punctures, while not completely airtight, sufficiently protect the vessel's contents from contamination and allow the gases produced by fermentation to evacuate from the vessel as the pressure rises and the balloon inflates.

Co-fermentation

Co-fermentation is the practice in winemaking of fermenting two or more grape varieties at the same time when producing a wine. This differs from the more common practice of blending separate wine components into a cuvée after fermentation. While co-fermentation in principle could be practiced for any mixture of grape varieties, it is today more common for red wines produced from a mixture of red grape varieties and a smaller proportion of white grape varieties.

Co-fermentation is an old practice going back to the now uncommon practice of having field blends (mixed plantations of varieties) in vineyards, and the previous practice in some regions (such as Rioja and Tuscany) of using a small proportion of white grapes to "soften" some red wines which tended to have harsh tannins when produced with the winemaking methods of the time. It is believed that the practice may also have been adopted because it was found empirically to give deeper and better colour to wines, which is due to improved co-pigmentation resulting from some components in white grapes.

Use Today

The only classical Old World wine region where co-fermentation is still widely practiced is now the Côte-Rôtie appellation of northern Rhône, while the use of white varieties in red Rioja and Tuscany wine has more or less disappeared. In Côte-Rôtie, the red variety Syrah and the aromatic white variety Viognier (up to 20% is allowed, but 5-10% is more common) must be cofermented, if Viognier is used. The reason why Viognier has been kept in Côte-Rôtie (while for example the white grapes Marsanne and Roussanne are hardly found any more in red Hermitage or other red Rhône wines where they are allowed) is that it adds signature floral aromas to the wines. The popularity of Côte-Rôtie has led to New World interpretations of this blend, most notably Australian Shiraz-Viognier blends, which are also produced by co-fermentation.

The reason why co-fermentation is not more widely practiced is that it "locks in" a certain blend already at the start of the fermentation, which gives the winemaker less possibility to adjust the blend after fermentation.

Co-fermentation is also performed in situations where Field blend varietals are indistinguishable from each other, thus necessitating co-fermentation.

References

- Hui YH, Meunier-Goddik L, Josephsen J, Nip WK, Stanfield PS (2004). Handbook of Food and Beverage Fermentation Technology. CRC Press. pp. 27 and passim. ISBN 978-0-8247-5122-7.
- Chandan, Ramesh C.; Kilara, Arun (22 December 2010). Dairy Ingredients for Food Processing. John Wiley & Sons. pp. 1–. ISBN 978-0-470-95912-1.
- Peters, Pam (2004). The Cambridge Guide to English Usage. Cambridge: Cambridge University Press, pp. 587–588, ISBN 052162181X.
- Don Tribby. Yoghurt. Chapter 8 in The Sensory Evaluation of Dairy Products. Eds. Stephanie Clark, et al. Springer Science & Business Media, 2009 ISBN 9780387774084 Page 191
- Coyle, L. Patrick (1982). The World Encyclopedia of Food. Facts On File Inc. p. 763. ISBN 978-0-87196-417-5. Retrieved 11 August 2009.
- Hui, ed. Ramesh C. Chandan, associate editors, Charles H. White, Arun Kilara, Y. H. (2006). Manufacturing yogurt and fermented milks (1. ed.). Ames (Iowa): Blackwell. p. 364. ISBN 9780813823041.
- A. Garrido Fernandez; M.J. Fernandez-Diez; M.R. Adams (31 July 1997), Table Olives: Production and Processing, Springer, pp. 23–45, ISBN 978-0-412-71810-6
- Alfred W. Crosby (2003). The Columbian Exchange: Biological and Cultural Consequencies of 1492. Santa Barbara, CA: Praeger. p. 73. ISBN 978-0-27598-092-4.
- Cooper, John (1993). Eat and Be Satisfied: A Social History of Jewish Food. New Jersey: Jason Aronson Inc. pp. 4–9. ISBN 0-87668-316-2.
- Enciclopedia Universal Europeo Americana. Volume 15. Madrid. 1981. Espasa-Calpe S.A. ISBN 84-239-4500-6 (Complete Encyclopedia) and ISBN 84-239-4515-4.
- Patrick F. Fox; P. F. Fox (2000). Fundamentals of cheese science. Springer, 2000. p. 388. ISBN 9780834212602. Retrieved March 21, 2011.
- Patrick F. Fox; P. F. Fox (1999-02-28). Cheese: chemistry, physics and microbiology, Volume 1. Springer, 1999. p. 1. ISBN 9780834213388. Retrieved March 23, 2011.
- Dilip K. Arora, Libero Ajello, K. G. Mukerji, Handbook of Applied Mycology: Foods and Feeds, Volume 3, CRC Press, 1991, ISBN 0-8247-8491-X.

Permissions

All chapters in this book are published with permission under the Creative Commons Attribution Share Alike License or equivalent. Every chapter published in this book has been scrutinized by our experts. Their significance has been extensively debated. The topics covered herein carry significant information for a comprehensive understanding. They may even be implemented as practical applications or may be referred to as a beginning point for further studies.

We would like to thank the editorial team for lending their expertise to make the book truly unique. They have played a crucial role in the development of this book. Without their invaluable contributions this book wouldn't have been possible. They have made vital efforts to compile up to date information on the varied aspects of this subject to make this book a valuable addition to the collection of many professionals and students.

This book was conceptualized with the vision of imparting up-to-date and integrated information in this field. To ensure the same, a matchless editorial board was set up. Every individual on the board went through rigorous rounds of assessment to prove their worth. After which they invested a large part of their time researching and compiling the most relevant data for our readers.

The editorial board has been involved in producing this book since its inception. They have spent rigorous hours researching and exploring the diverse topics which have resulted in the successful publishing of this book. They have passed on their knowledge of decades through this book. To expedite this challenging task, the publisher supported the team at every step. A small team of assistant editors was also appointed to further simplify the editing procedure and attain best results for the readers.

Apart from the editorial board, the designing team has also invested a significant amount of their time in understanding the subject and creating the most relevant covers. They scrutinized every image to scout for the most suitable representation of the subject and create an appropriate cover for the book.

The publishing team has been an ardent support to the editorial, designing and production team. Their endless efforts to recruit the best for this project, has resulted in the accomplishment of this book. They are a veteran in the field of academics and their pool of knowledge is as vast as their experience in printing. Their expertise and guidance has proved useful at every step. Their uncompromising quality standards have made this book an exceptional effort. Their encouragement from time to time has been an inspiration for everyone.

The publisher and the editorial board hope that this book will prove to be a valuable piece of knowledge for students, practitioners and scholars across the globe.

Index

www.ingramcontent.com/pod-product-compliance
Lightning Source LLC
Chambersburg PA
CBHW061934190326
41458CB00009B/2738